T0230451

SpringerBriefs in Computer Science

SpringerBriefs present concise summaries of cutting-edge research and practical applications across a wide spectrum of fields. Featuring compact volumes of 50 to 125 pages, the series covers a range of content from professional to academic.

Typical topics might include:

- A timely report of state-of-the art analytical techniques
- A bridge between new research results, as published in journal articles, and a contextual literature review
- A snapshot of a hot or emerging topic
- An in-depth case study or clinical example
- A presentation of core concepts that students must understand in order to make independent contributions

Briefs allow authors to present their ideas and readers to absorb them with minimal time investment. Briefs will be published as part of Springer's eBook collection, with millions of users worldwide. In addition, Briefs will be available for individual print and electronic purchase. Briefs are characterized by fast, global electronic dissemination, standard publishing contracts, easy-to-use manuscript preparation and formatting guidelines, and expedited production schedules. We aim for publication 8–12 weeks after acceptance. Both solicited and unsolicited manuscripts are considered for publication in this series.

**Indexing: This series is indexed in Scopus, Ei-Compendex, and zbMATH **

Blaž Škrlj

From Unimodal to Multimodal Machine Learning

An Overview

 Springer

Blaž Škrlj
Jožef Stefan Institute
Ljubljana, Slovenia

ISSN 2191-5768 ISSN 2191-5776 (electronic)
SpringerBriefs in Computer Science
ISBN 978-3-031-57015-5 ISBN 978-3-031-57016-2 (eBook)
https://doi.org/10.1007/978-3-031-57016-2

This Springer imprint is published by the registered company Springer Nature Switzerland AG
The registered company address is: Gewerbestrasse 11, 6330 Cham, Switzerland

Paper in this product is recyclable.

To Oliver, for who you'll become.

Preface

Machine learning has in the recent years become a ubiquitous technology in many areas of science and industry, enabling many elusive use cases where human intuition alone is not sufficient. Multimodal machine learning, which involves processing and learning from data across multiple modalities, has opened up new possibilities in a wide range of applications, including speech recognition, natural language processing and image recognition.

This book focuses on the internal representations used by multimodal machine learning algorithms and how they can be effectively combined to develop powerful machine learning systems that can handle complex real-world scenarios. We provide an overview of the field, including different modalities used in multimodal machine learning and the challenges and opportunities associated with processing and learning from data across multiple modalities. We then explore different approaches to solving the challenge of combining information from different modalities, including different types of representation fusion and mixture-of-experts-type of neural network architectures. We discuss the advantages and disadvantages of each approach and provide examples of their applications in real-world scenarios. Another critical aspect of multimodal machine learning is how to evaluate the performance of models. Traditional evaluation metrics, such as accuracy and precision are often insufficient in the context of multimodal machine learning, as they do not take into account the interactions between different modalities. Therefore, examination of novel evaluation scenarios, such as cross-modal retrieval and cross-modal classification, that take into account the interactions between different modalities is an emerging research area. Throughout the book, we provide examples and illustrations to help readers understand the concepts and methods discussed.

Multimodal machine learning is an exciting and rapidly evolving field that has the potential to revolutionize many areas of science and industry. By effectively combining information from different modalities, researchers can develop powerful machine learning algorithms that can handle complex real-world scenarios. This book provides a valuable resource for researchers and practitioners interested in this field.

Ljubljana, Slovenia   Blaž Škrlj
March 2024

Acknowledgements

This work was partially funded by ARIS project with reference number J4-4555. This work was also partially financed by financial support from the Slovenian Research Agency for research core funding for the programme Knowledge Technologies (No. P2-0103).

Contents

Acronyms

ML	Machine Learning
GAT	Graph Attention Networks
GNN	Graph Neural Network
LSTM	Long-Short Term Memory
MF	Matrix Factorization
FFN	Feedforward neural network
MoE	Mixture of Experts
GPT	Generative Pre-trained Transformer
BOW	Bag-of-words
cBOW	Continuous Bag-of-words
TF-IDF	Term frequency - inverse document frequency
LLM	Large language model
NetMF	Network embedding as Matrix Factorization
ReLU	rectified linear unit
IOU	Intersection over union
GAN	Generative Adversarial Network

Part I
Introduction

This book delves into the core focus of modern machine learning, which is the ability to automatically extract patterns and encode data internally into a form that enables efficient comparisons and inference. It emphasizes the significance of understanding the necessary internal representations, their differences, and the trade-offs they offer, essential for building successful real-life machine learning systems.

This work serves as a reference for machine learning practitioners who need to expand their knowledge to incorporate additional information in everyday problems or research. The following chapters discuss in detail the historical perspective of machine learning and artificial intelligence, the development of unimodal and multimodal approaches, and the first unimodal algorithms that operate with text, images, and graphs, followed by multimodal algorithms. Each chapter is structured as a collection of sections, including Method spotlights, that offer an overview of the topic or a "deep dive" into the selected method. The technical sections are intended for interested readers familiar with the basics of linear algebra and who would like to understand seminal work from different domains on a deeper level.

The first chapter discusses the origins of machine learning and artificial intelligence, aligned with the main problems that persist today, such as causal reasoning, and the research conducted in the era of very limited hardware and software. It highlights key components of learning and seminal ideas that persisted throughout past decades, including the notion of learning, robotics-based research, capabilities of reinforcement learning, and neural network-based approaches, the dominant technology of machine learning since the 2010s.

Chapter 1
A Brief Overview of Machine Learning

1.1 On Artificial Intelligence and Machine Learning

Early approaches and ideas that attempted to disseminate the notion of "automated" learning from data date back to 1950s. One of the first concrete formalizations of a system capable of learning, inspired by human neurons, was created by Donald Hebb (The Organization of Behavior) [52]. This work focuses on a cell model (a surprisingly aligned one to contemporary neural networks for that matter). Not long after his work, Alan Turing published his seminal paper entitled "Computing Machinery and Intelligence" [132]. This paper introduces the renowned **Turing test** and, with it, jump-starts the notion of artificial intelligence as we know it today. The term, however, was coined on a workshop proposal in 1956 by Claude Shannon, Marvin Minsky, John McCarthy, and Nathaniel Rochester. This event is known as the founding event of the AI field.

Another seminal work that is of particular relevance to the dominant branch of methods nowadays (**neural networks**) is Frank Rosenblatt's perceptron paper [110]. In it, he conceptualized the architecture of an artificial neuron and pinpointed the training procedure. He introduced the notions of *learning rate* and differentiable weight spaces alongside an optimization algorithm (steepest decent) that enabled finding suitable sets of weights for particular training data. The term "machine learning," however, was coined at the end of 1950s by Arthur Samuel [115], who published a seminal work on an artificial system capable of playing the game of checkers (self-improving agent). His system considered the minimax algorithm with alpha–beta pruning, a technique that gave rise to superhuman game-playing engines (Chess, go, and similar). Samuel also pushed a substantial update to his early work 8 years after the first paper, where he discussed notions such as long-term strategy, learning slowness, and non-linear approximation of states – many of these problems remain open for hard real-life problems nowadays [116].

Gradual progress has been observed throughout the 1960s, with systems such as BOXES [86] and similar, that explored the limits of adaptive learning. This decade also gave rise to first **chatbots**. One of the first such systems (after the thought experiment of Turing) was Eliza[1] [140]. This was one of the first systems that explored the notion of a human–computer dialogue, including possible implementations of it. This year also marks the debut of humanoid robotics; Shakey, the Stanford labs' robot, was actively developed from 1966 to 1972 (Charles Rosen, Nils Nilsson, and Peter Hart led the project) and could perform tasks that required planning, route finding, and the rearranging of simple objects. The robot greatly influenced modern robotics and AI techniques used nowadays [72]. Another seminal achievement in this decade was the introduction of *nearest neighbour* algorithm [25] that saw interesting applications in route planning.

Throughout 1970s, there was a shift from neural network-based methods to symbolic AI – branch of approaches that leverage human-obtained background knowledge and aim to derive logical rules. One of the influential works of this era was *perceptrons* [90], a paper that mapped out limitations of neural network-based approaches. The remainder of this decade saw funding decrease. However, considerable improvements to existing methods were introduced already around the 1980s, when a general rule learning system was proposed by Dejong and Mooney [27]. The proposed system could discard irrelevant data and produce general explanations of training data (rules). Further, this era also saw considerable improvements when it came to computer speech. Systems capable of pronouncing English were already considered by Sejnowski and Rosenberg [118]. Their neural network was comprised of 203 inputs and 80 hidden units, amounting to 18,629 weights. Throughout the 1990s, convolutional neural networks were introduced and applied to real-life use cases like document recognition [75] and time series analysis. Furthermore, the notion of Q-learning, the basis for modern reinforcement learning, was also introduced in this period [139]. Note that Q-learning was defined by Walkins in 1989 ("a simple way for agents to learn how to act optimally in controlled Markovian domains"). By the end of the 1990s, Sepp Hochreiter and Jürgen Schmidhuber proposed **recurrent** neural networks, an architecture that enabled the processing of longer sequential data [55]. Variations of this architecture are used nowadays in handheld devices for speech recognition, as well as building blocks for reinforcement learning systems. This decade also gave rise to MNIST, a widely recognized data set for recognition of handheld digits (used in many research works nowadays) [74].

1.2   Contemporary Machine Learning

We proceed by discussing the progress in the last 20 years of research. The notion of "contemporary" here merely implies many of these methods remain actively

---

[1] https://web.njit.edu/~ronkowit/eliza.html

developed, as even though decades old, many ideas mentioned in the previous section remain lively research areas on their own. In the early 2000s, the first neural language model was proposed by Montreal researchers [9]. This system was neural network-based and is in many aspects similar to chatbots that have been becoming mainstream since the early 2020s. Interestingly, this decade (the 2000s) also marks many contributions that lie more on the engineering side of research – researchers of that era realized that without high-quality software to support ever-faster learning/research cycles, progress could be hindered. One of the first such libraries was Torch [24] – a collection of C-based routines that enabled reusable neural network-based experiments. This decade also saw the introduction of ImageNet [29], a large-scale database of images that enabled many subsequent advancements in deep learning. This decade also saw interesting developments in the field of *recommender systems* – systems that entail a learning component and are built to help prioritize items of various sorts (e.g., movie titles). The Netflix prize was one of the first larger shared tasks on this topic and offered many interesting approaches for solving this task (collaborative filtering).[2] The subsequent decade (2010s) saw some substantial improvements in the field of deep learning – a term coined around 2006 [53]. The early 2010s saw many improvements at the level of *implementation* of neural networks. In particular, specialized hardware such as GPUs (Graphics Processing Units, initially used mostly in the game industry) started to become used for an increasing number of state-of-the-art neural network-based approaches [23].

One of the landmark achievements of this decade is the architecture that won the ImageNet challenge [70]. This work is also renowned as the pivot point for the research field of deep learning to start focusing more on the implementations, as with better implementations, deeper neural networks are possible; so far, deeper networks, on many occasions, offer supreme results to more shallow counterparts. In the area of neural language processing, a seminal paper by Google researchers proposes *word2vec*, a method for obtaining distributed representations of words (embeddings) [87]. In the same time period, Generative Adversarial Networks (GANs) were introduced [43]. This method enabled the automated generation of patterns (initially images) by adopting a two-network approach; discriminator and generator networks jointly either learn to discriminate artificial from real or generate the data itself. Joint training gives rise to generator networks that can output novel/derived representations and, as such, generate interesting variations of the input. Similarly, work that touches the topic of generative modelling from this era relates to generative variational autoencoders [64] – this idea explored the notion of variational inference for the task of autoencoding and, with it, simplifies the generative process (traversal of probabilistic, defined space for generative step).

Researchers at DeepMind proposed a system that achieved superhuman performance in a collection of challenging games, including Chess, Go, and Shogi [120]. The system, termed AlphaZero, achieved this feat via self-play. This paradigm

---

[2] https://web.archive.org/web/20200510213032/https://www.netflixprize.com/assets/rules.pdf

focuses on pitting the system against itself with the aim of gradual improvement. Compared to existing systems, e.g., chess, where substantial knowledge about previous games is encoded, AlphaZero learned to play better without any human intervention (apart from rule specification). The last few years have seen considerable improvements in the field of natural language processing (and, more recently, multimodal learning). The *transformer*-based neural network architecture offers a more scalable alternative to previously adopted recurrent variants [133]. Attention layers can be understood as learnable pairwise interaction mappings within a given context window and enable the construction of systems comprised of billions of parameters (with correspondingly good performance). Architectural improvements, combined with increasingly available text data, enabled the era of *language models*. These systems commonly optimize for surprisingly simple objectives (e.g., predict the missing/next token) yet are consistently shown to offer good performance across a plethora of tasks, from language understanding and classification to question answering. One of the methods that offered across-the-board improvements and is publicly available is BERT [30].

As training such systems from scratch is computationally demanding, and as such limited to entities with enough compute, model *fine-tuning* became a prominent research niche on its own. This procedure, instead of training the whole model from scratch, re-uses the existing model as the "base model" and, with only a few passes across specific data, enables substantial improvements in performance on much less hardware. There exist many variations of BERT, tuned for different domains – from biology and finance to engineering. The most recent branch of improvements in this field concerns *generative* language modelling – the tasks where a system is asked to predict the next token. By being able to do so, systems can also *generate* new text (or some sequential input). The Generative Pre-trained Transformer (GPT) family of models represents current state of the art [98]. It was recently shown that this type of architecture if pre-trained on enough data possesses zero-shot capabilities; they are able to solve tasks previously unseen (in exact form) with minimal additional training (or even no fine-tuning) [14]. This branch of methods gave rise to commercial chatbots that are being widely adopted (e.g., ChatGPT, Claude, and similar).

This chapter of the book attempted to equip the reader with some high-level intuition about progress in the field of machine learning (and artificial intelligence). We refer the more interested reader to recent, more history-focused works [2, 5, 112, 113] that provide a more in-depth understanding of key pivot points and achievements in the past 70+ years.

Chapter 2
Data Modalities and Representation Learning

2.1 Common Data Modalities

We proceed with an overview of the main modalities considered throughout this book. For each considered modality, we discuss where it is commonly present (main use cases) and some of the main approaches that are considered state-of-the-art.

2.1.1 Learning from Tabular Data

One of the most commonly considered learning settings considers *tabular data* (the scheme in Fig. 2.1). This type of problem can be understood as learning mappings between matrix-based inputs (tables) and a dedicated output space. The output space can be categorical or numeric (classification or regression). Furthermore, one or more classes/targets can be predicted simultaneously, giving rise to computationally more efficient learning that scales better for more complex learning scenarios. Apart from classification, a commonly addressed task when considering tabular data is also *regression* – the task of predicting continuous outputs (e.g., precipitation based on environmental data).

Algorithms for learning from tabular data adhere to many different learning principles, as this data source is among the most studied. In particular, tree-based methods that emerged at the end of 1980s [96, 97] and, in ensemble form, in subsequent decades remain state-of-the-art for many real-life data sets. One of the first ensemble-based tree algorithms was Random Forests [13]. This approach considers a collection of trees trained on different sub-spaces of the data; trees, in the end, vote to obtain the final prediction. An alternative branch of tree ensembles that is also competitive on many shared tasks is gradient-boosted trees. An example implementation, XGBoost, constructs trees in an iterative manner, each iteration accounting for mistakes of the models already considered (boosting) [19]. Further,

© The Author(s), under exclusive license to Springer Nature Switzerland AG 2024
B. Škrlj, *From Unimodal to Multimodal Machine Learning*, SpringerBriefs in Computer Science, https://doi.org/10.1007/978-3-031-57016-2_2

Fig. 2.1 Schematic overview
of learning from tabular data.
Input feature space
(differently typed columns) is
used as input to learn a
mapping to a tabular output
space (multiple targets,
commonly of the same type)

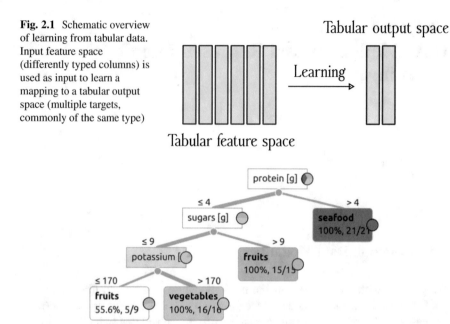

Fig. 2.2 An example decision tree depicting classification of food types based on nutrient presence. Visualization and data obtained with Orange3 toolkit [28]. Tabular data sample considered as input consists of three features (protein, sugar, and potassium mass)

the main implementation of boosting ideas focusing on categorical inputs is known as CatBoost [94]. A real-life example of a decision tree is shown in Fig. 2.2. Here, the feature representing protein mass (in grams) was recognized as the most important one (first split), followed by the presence of sugars and potassium mass. Note that splits are numeric – thresholds are considered instead of particular categorical values.

The above examples illustrate the prominent branch of methods (tree-based methods) for learning from tabular data. Alternative approaches, such as neural network-based algorithms and similar, are also becoming competitive on many benchmarks. A recent example of a neural network-based approach targeted at tabular data learning is TabNet [3]. Apart from the classification and regression tasks discussed above, tabular data are also commonly used as input for clustering-based analysis and feature ranking – the process of identifying relevant features in a data set with the aim of better understanding the main patterns.

2.1.2   Learning from Text Data

Learning from text-based data is becoming ubiquitous across many fields, mostly due to the recent advancements in the area of neural language models [88] – deep learning-based approaches that operate by learning to predict subsequent tokens in a given token stream. Throughout this work, we consider two main approaches to learning from texts: deep learning-based approaches and symbolic approaches (Bag-of-Word-based approaches).

Deep learning approaches for text classification and other tasks have emerged as the dominant paradigm in the last 10 years (2010s). The first wave of methods that enabled text-based learning were *recurrent neural networks*. This branch of networks became the *de facto* method around the time of the seminal paper about long- and short-term memory in this type of network, the LSTM (Long Short-Term Memory) neural networks. The proposed architecture addressed the issue of *vanishing gradient* problem – a phenomenon discovered already in the 1990s [54]. The problem describes a scenario common to "generic" recurrent neural networks, where intermediary gradients tend to vanish (or explode) during optimization due to recurrent connections in the network. The issue was prohibitive for scaling recurrent neural networks to a more significant extent. The LSTM networks overcame the issue by adopting architectural changes to each recurrent layer introduced, enabling state-of-the-art sequence labelling results with recurrent neural networks. An alternative approach to using LSTMs that has proven faster is Gated Recurrent Units (GRUs) [21], introduced a few years after the LSTMs.

Around the same time period, representation learning techniques for mapping either tokens (words) or whole documents to latent vector spaces emerged. Prominent examples include *word2vec* [87] – a highly efficient algorithm for producing representations of words, and *doc2vec*, its extension to representing whole documents [73]. The two tools enabled a collection of applications, ranging from semantic search (vector similarity-based retrieval of documents) to classification and topic modelling [62]. More recent advancements in the area of efficient representation learning of tokens include the inclusion of sub-word information for increased accuracy [10]. Representations of tokens could be pre-trained and publicly offered for more than 150 languages, enabling almost zero-friction deployment for specific use cases where computation is limited [45]. An example of distances between token representations (in 3D) is shown in Fig. 2.3. This branch of methods, albeit highly scalable, however, lacks the precise notion of *context*. Obtaining a static representation for, e.g., a token "bank" might imply there is only one meaning to it; this is not the case (e.g., river bank and national bank are very different). Issues like this one eventually resulted in the most recent branch of models that revolve around *language modelling*. Here, larger neural network-based models (initially recurrent, more recently transformer-based) are trained in a generative manner. For large amounts of text, their goal is to predict subsequent tokens. As such, this, with enough iterations, results in language models developing internal *contextual representations*. This property enables more fine-grained task solving and paved the way to human-level performance across many tasks that concern language understanding and even reasoning.

Approaches such as BERT [30] and other transformer-based architectures [133] dominate public leader-boards for the adopted benchmarks such as GLUE [137] and more recently superGLUE [136]; instead of being able to utilize pre-trained token representations to build more complex representations (e.g., *doc2vec*-type

Fig. 2.3 Schematic representation of semantic distance in three dimensions. This illustrative example shows that the distance between the concepts of "man" and "woman" is approximately the same as the one between "king" and "queen"

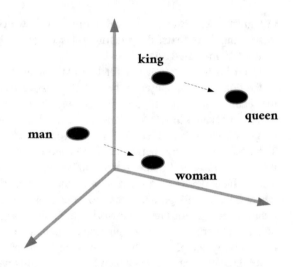

of approaches do), language model approaches rely on the notion of *pre-trained* models. At least two distinct tasks are commonplace when considering language model-based learning: model pre-training, the process of generative modelling on large amounts of text, and model fine-tuning, the process of adapting the pre-trained model for a specific task. As pre-trained models are becoming increasingly generic, *repositories* of such models are becoming the *de facto* storage and source of such models for many researchers, as pre-training larger neural networks is mostly out of reach for the most medium to small-scale labs. Projects such as HuggingFace [142] are becoming increasingly valuable as they facilitate the fast iterations with many ready-to-go models. As language models effectively hold the representations spread across the whole architecture, projects such as *sentencebert* emerged, enabling across-layer aggregations of weights with the aim to obtain *contextual document representations* [102].

The alternative form of document and text representation is commonly called the Bag-of-Words branch of approaches. This type of approach was the common solution in the 1990s and 2000s, even though even earlier approaches exist [59, 152]. Here, *sparse* representations of data are obtained by mapping and re-weighting the presence of tokens at, e.g., the document level. For example, in the sentence "This cat eats only fresh food," a generic BoW algorithm would identify tokens such as "cat" and "food" as the relevant ones for building the representation. The two tokens would represent two features (=dimensions). Note that in this case, dimensions have an explicit meaning and are thus *explainable* (compared to neural approaches, where this is not the case). Values for the features can be obtained by either counting occurrences of the token, accounting only for presence, or adopting one of the many possible heuristics to re-weight the values based on global counts. The trade-offs associated with this type of representation lie in the fact that it could be too coarse-grained for competitive performance on tasks where language models currently dominate. On the other hand, their sparse nature and low data requirements

Fig. 2.4 Schematic overview of learning from images (classification). A collection of images (coloured input blocks in the scheme) is used as input, and the output is structured (mostly tabular) space, for example, object types

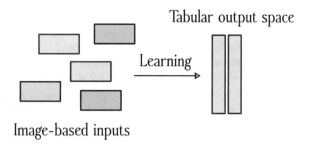

render this branch of methods useful in scenarios where computing and/or data are limited. Further, for document retrieval tasks, hybrid approaches are a promising research venue [82].

2.1.3 Learning from Image-Based Data

One of the more computationally demanding input types are *images* (the scheme in Fig. 2.4). Learning from images has undergone one of the most apparent paradigm shifts in the last 15 years – from initially using hand-crafted approaches; the focus has shifted almost exclusively to deep learning-based approaches. This branch of approaches enabled real-time image segmentation, classification, generation, and similar tasks. Furthermore, models aimed at understanding video-based inputs (ordered sequences of images) have also emerged as a separate research endeavour. We continue with a brief overview of some of the key developments in the area of deep learning for computer vision.

One of the first neural network architectures that dominated public machine learning challenges was AlexNet [70], an architecture that introduced the notion of ReLU activation to this type of model. It is comprised of five convolutional layers and three fully connected layers and, further, uses a GPU-parallel training regime to accommodate the use of two GPUs during training. Following this advancement, a batch of deeper models emerged. The Inception V1 architecture [125] consists of 22 hidden layers organized in smaller functional groups. The authors have shown that using the RMSProp optimization algorithm (adaptive learning rate; see [111] for an overview), alongside batch normalization, offered additional improvements. The trend of constructing even deeper neural network architectures continued with VGGNet [121], an architecture that used up to 19 hidden layers. This architecture uses smaller filters (3×3), enabling deeper neural networks to be processed in reasonable times. It also includes intermediary pooling layers – layers that appear in every convolutional layer and are responsible for intermediary information aggregation.

One of the prominent neural network architectures that remain used to this date is ResNets [50]. This model was designed with scalability in mind – its *skip connections* enables the pass-over of information across multiple layers. This branch of models also employs batch normalization. ResNet architectures that are openly available range from 18 layers to more than 1000 layers. With larger neural networks, the training and inference can become prohibitively expensive. This is one of the reasons for the Xception architecture, an update of the aforementioned Inception V1, built with better scalability in mind. Xception introduces depthwise (separable) convolutions, a type of layer that enables separation of spatial and non-spatial correlations in images; it first captures cross-feature correlations, followed by processing of spatial correlations. One of the latest advancements in the field is the ResNeXt architecture [145]. This paper introduces "cardinality" (the size of the set of transformations) as an essential factor in addition to the dimensions of depth and width; overall, the block-like structure of the networks facilitates hyperparameter optimization (a smaller architecture search is required). Around the same time, image segmentation was also substantially improved due to the introduction of YOLO-based architectures [100]. This state-of-the-art approach to image segmentation (the eighth version is available at the time of writing) enabled parsing of around 50 frames per second. The main improvement came from formulating object detection as a regression problem to spatially separated bounding boxes and associated class probabilities. This implies a single neural network predicts bounding boxes and class probabilities directly from full images in one evaluation; this insight enables end-to-end learning (instead of multi-approach pipelines).

The field of generative image modelling has also been under active research – the paper by Goodfellow et al. [43] on generative adversarial neural networks showcased one of the ways to train high-quality generative models with the aid of a model that is trained to discriminate (adversarial role to the generator network). Even more recent advancements indicate that self-supervised learning for images is also a viable research endeavour with many applications. Masked autoencoders, an architecture that attempts to learn the masked pixels (similar to masked language models), were shown to perform this task well, enabling self-supervised training for image-based data sets – lack of labelled data in this domain can be prohibitive for some use cases [49].

2.1.4   Learning from Graph-Based Data

Many forms of data are inherently relational. In such cases, *graphs* (the scheme in Fig. 2.5) are a suitable abstraction for representing the data sets of interest, resulting in *graph-based* machine learning. Examples of graph-based data sets include street/traffic data, biological interaction data (e.g., protein interactions), and social network-based data [143]. Graph-based machine learning attempts to exploit the information stored in *the structure* of the relation networks. Apart from structure alone, methods of this kind are also capable of accounting for attributes – properties of nodes. Information propagation algorithms [155] propagate label information via

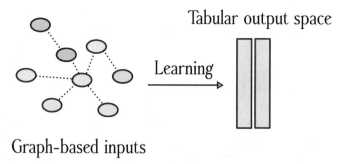

Tabular output space

Learning

Graph-based inputs

Fig. 2.5 Schematic overview of learning from graph-based data. Compared to images and texts, learning can take place within a graph (as shown in this image) or across many graphs

nodes' neighbours until all nodes are labelled. These algorithms learn in an *end-to-end* manner, meaning that no intermediary representation of a network is first obtained and subsequently used for training, e.g., a classifier.

Another class of node classification algorithms learns node labels from node embeddings, i.e., node representations in vector form [26]. Here, the whole network is first transformed into an information-rich, compact low-dimensional representation (a dense matrix). This representation serves as an input to a plethora of more general machine learning approaches that can be used for node classification.

We distinguish between two main branches of embedding-based learning algorithms, discussed next: graph neural networks and random walk-based learners. Graph neural networks (GNNs), introduced in recent years, attempt to incorporate a given network's adjacency structure as new neural network layers. Among the first such approaches were the graph convolutional networks (GCNs) [65], their generalization with the attention mechanism [134], and the more recent isomorphism-based variants with provable properties [146]. Treating the adjacency structure as a neural network has also shown promising results [48]. The key characteristic of this branch of methods is their capability to account for *node features* by multiplication of the normalized adjacency matrix as part of a special layer during learning from features. On the other hand, if node features are not available, which is the case with the majority of freely available public data sets, more optimized methods focused on *structure-based learning* are preferred. For example, the LINE algorithm [127] uses the network's *eigendecomposition* in order to learn a low-dimensional network representation, e.g., a representation of the network's nodes in 128 dimensions instead of the dimension that matches the number of nodes. Approaches that use random walks to sample the network include DeepWalk [92] and its generalization node2vec [46].

It was recently proven that DeepWalk, node2vec, and LINE can be reformulated as implicit matrix factorization [95]. Furthermore, approaches such as struc2vec [106] demonstrated how more complex, multilayer structures can be compressed into node representations for better performance. Despite many promising approaches developed, a recent extensive evaluation of network embedding techniques [44] suggests that node2vec [46] remains one of the best embedding approaches for the task of **structural** node classification.

2.2   Representation Learning

Having discussed different modalities of potential interests, this chapter concludes with an overview of the notion of *representation learning*, one of the paradigms that enables the unification of modalities and will be of focus throughout this book.

2.2.1   What Is Representation Learning?

Representation learning corresponds to the process of *deriving* (mostly) real-valued data from initial data of a given modality. The learned representations are further characterized as being a *latent space* – the dimensions (columns) in the obtained data sets of representations do not correspond to any human-understandable trait but represent "coordinates" of a higher dimensional space within the derived representation of the object of interest resides. Contemporary representation learning is most commonly associated with neural network-based learning, even though it is not exclusive to this branch of methods. Neural networks are suitable representation learners because their internal representations are dense real-valued vectors – a data structure that directly, or via aggregation, corresponds to final learned representations. Representation learning can be linear or non-linear. Linear representation learning examples include representations obtained via singular value decomposition or similar matrix factorization procedures, whereas non-linear representations mostly result from neural network-based learning. Conceptual overview of representation learning can be seen in Fig. 2.6. Many machine learning algorithms and approaches can be characterized as representation learners; the key difference lies in the utility of the representations. With the advent of neural network-based methods, intermediary real-valued representations (embeddings) are already suitable inputs for downstream learning tasks; instead of being used in an end-to-end manner to solve specific tasks, learned representations can be *re-used*. Note that some approaches, even though trained end to end, enable similar transfer of knowledge (language models). Examples of representation learning are discussed next.

2.2.2   Examples of Representation Learning

The main advantage of representation learning is that once the computationally expensive process of obtaining the representations is concluded, the results can

Fig. 2.6 Conceptual overview of representation learning. Different types of data (coloured input blocks) are transformed into the same format (numeric representations)

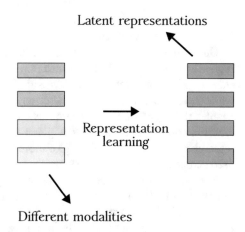

be used as inputs for multiple downstream tasks. This approach is particularly useful for tasks where data is scarce or expensive to obtain. For example, node representation learning is an important aspect of graph analysis, and one example of a representation learning algorithm for this purpose is DeepWalk [92]. DeepWalk produces k-dimensional representations of nodes in a graph, thereby enabling the efficient learning of node embeddings. The DeepWalk algorithm works by generating random walks on the graph and using the Skip-Gram model to obtain the node embeddings (inspired by word2vec [87]). Although the initial process of obtaining the representations can be time-consuming, the results can be used as inputs for various tasks such as node classification, link prediction, outlier detection, and similarity search.

Node classification is a common task in graph analysis that involves predicting the labels of unlabeled nodes in a graph. Logistic regression can be used to perform the final inference in node classification, given the learned representations and the corresponding labels. Further, link prediction is another task that aims to predict the existence of edges between nodes in a graph. Outlier detection is another task commonly tackled in graph analysis. The learned representations can be used to identify nodes that are different from the others in the graph, indicating that they might be outliers. Finally, similarity search is also an important task in graph analysis. The learned representations can be used to compute dot products or other similarity measures, which can be used to assess the similarity between nodes in the graph. For example, the cosine similarity measure can be used to compare two nodes based on the angle between their corresponding representations. If the angle is small, the nodes are similar, while if the angle is large, the nodes are structurally different. The DeepWalk algorithm is just one example of a representation learning algorithm for node embeddings. Other algorithms, such as node2vec [46] and GraphSAGE [48], have also been developed to address various challenges in node representation learning. Node2vec is an extension of DeepWalk that uses a biased random walk strategy to generate node embeddings, while GraphSAGE is a neural network-based approach that aggregates the local neighbourhood information to generate node embeddings.

One of the key properties of representation learning is reusability of the learned representations for multiple downstream tasks. By using pre-trained models as a starting point, researchers can save significant computational resources and time while still achieving high performance on a variety of tasks. Moreover, this approach enables the separation of concerns when it comes to learning more foundational representations versus solving a given task at hand. This separation can help researchers focus on the learning of more generalizable representations that can be used across different problem domains.

However, not all methods adhere to the standard paradigm of learning representation plus downstream learning tasks. Large language models, for example, require dedicated approaches to extract meaningful representations from multiple layers of a neural network. These approaches enable the extraction of holistic representations from complex models, making them easier to use in downstream tasks. One such approach is the Sentence-BERT (SBERT) [102] algorithm, which has received significant attention in recent years. SBERT can create embeddings for entire sentences, producing a high-dimensional vector that conveys semantic meaning and can be used for a variety of tasks. This algorithm utilizes a Siamese network architecture to learn sentence embeddings by comparing pairs of sentences, thereby capturing the semantic similarity between them. SBERT has been shown to achieve state-of-the-art performance on a variety of natural language processing tasks, including sentence classification, semantic textual similarity, and question answering. By learning foundational representations that can be used across multiple tasks, the researchers can save significant time and resources. With the development of new techniques like SBERT, researchers are now able to extract meaningful representations from even the most complex models, making them easier to use in downstream tasks.

In recent years, learned representations have become more widely adopted, leading to the development of dedicated database systems that are specialized for storage and retrieval-based tasks with vector-based inputs. One example of such a system is Pinecone,[1] a vector database that has been developed to facilitate these tasks. Pinecone stores the learned representations and allows for efficient storage and retrieval, making it easier to use them for a variety of tasks.

---

[1] https://www.pinecone.io/learn/vector-database/

Part II
Unimodal Machine Learning

Machine learning is becoming ubiquitous throughout many areas of science and industry. The ability to automatically distil and associate patterns enabled many elusive use cases where human intuition alone is not sufficient (or as optimal). Each machine learning algorithm *encodes* the data internally into a form that enables efficient comparisons and inference. These internal *representations* are the core focus of this book. Understanding how and why such representations are necessary, how they differ, and what trade-offs they adhere to is key to building successful real-life machine learning systems.

Unimodal machine learning is a learning paradigm that focuses on representations of the same type. It is an approach that has been widely used in machine learning and AI research in recent years. Examples of unimodal machine learning include systems that can learn from text, images, or graphs. The single modality constraint ensures the relative simplicity of such approaches, which can be deployed and utilized at scale in real-life scenarios where a single modality is the only data source available.

Focusing on a single modality has been the primary focus of machine learning and AI communities in the past decades. One of the main reasons for this is the relative simplicity of approach development. Writing parsers and logic for handling dedicated modality required considerably less work than considering in-house solutions for handling multiple modalities. However, this changed with the advent of high-performance libraries that are the focus of methods in Chap. 3.

Text is one of the most ubiquitous modalities and is available in significant quantity. As text is one of the key means of human communication, linking knowledge via text-based interfaces is a promising research endeavor. Being able to process image-based data has a plethora of real-life use cases, such as traffic monitoring, pill anomaly detection, and biomedical image segmentation. Graph-based data are present when working with routing (map-based applications) and social networks. These unimodal approaches have enabled researchers to develop powerful machine learning algorithms that can handle these scenarios effectively.

Moreover, unimodal machine learning has allowed researchers to gain a better understanding of the intricacies of individual modalities. This approach has enabled the development of algorithms that can learn from large amounts of data and can make predictions based on this knowledge. In this sense, unimodal machine learning provides a solid foundation for the development of more complex and sophisticated machine learning algorithms. In conclusion, unimodal machine learning has been a focus of machine learning and AI communities in the past decades. This approach has enabled researchers to develop effective machine learning algorithms that can handle various scenarios. By focusing on a single modality, researchers have gained a better understanding of the intricacies of individual modalities, which has enabled the development of algorithms that can learn from large amounts of data and can make predictions based on this knowledge. Through this approach, researchers can build a solid foundation for the development of more complex and sophisticated machine learning algorithms.

Chapter 3
Learning from Text

3.1 Non-contextual Representation Learning

We begin with the overview of methods by discussing *non-contextual* representation learning algorithms. This branch of algorithms emerged as one of the first algorithms that scaled to very large quantities of text data while enabling *embedding*-based learning and inference. Compared to alternative algorithms discussed later in this chapter, embedding-based algorithms aim to compress concepts into $\mathbb{R}^k$. Once representations are obtained, downstream tasks can be solved by computing derived properties, e.g., linear combinations of existing (present) representations. Note that there are two possible interpretations of contextual representation learning – even though the algorithms discussed in subsequent chapters explicitly mention "context" as part of definitions, they produce non-contextual representations (static). Contextual representations produced by neural language models are obtained in a different manner, relying more on the dynamic nature of tokens (many meanings, context-dependent). We begin by discussing the reasoning underlying *word2vec*, one of the first main-stream embedding algorithms that enabled multiple of novel applications at scale.

3.1.1 Skip-Gram and Continuous Bag of Words

The two main ideas that have seen widespread adoption are the skip-gram and Continuous Bag-of-Words (cBOW) models [87]. Both paradigms enable efficient representation training while emphasizing different aspects of the input text (and how the target token is considered) (Fig. 3.1).

In the **cBOW** model, the target token's neighbourhood (context in a sentence) is used to predict the token itself. This includes tokens before and after the token

© The Author(s), under exclusive license to Springer Nature Switzerland AG 2024
B. Škrlj, *From Unimodal to Multimodal Machine Learning*, SpringerBriefs in
Computer Science, https://doi.org/10.1007/978-3-031-57016-2_3

Fig. 3.1 The cBOW model. A collection of tokens that constitute the target token's neighbourhood (token X is the target token) is used to obtain a representation that is used to predict the value of token X. The final result of cBOW training is thus token representations that are induced from their neighbourhoods

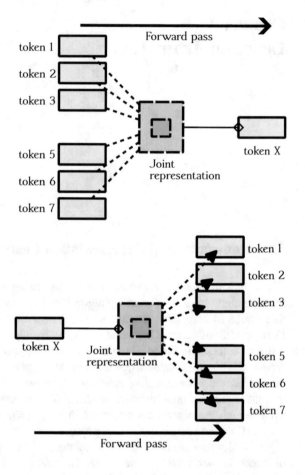

Fig. 3.2 The skip-gram model. Tokens are used to predict their neighbourhood (inverse problem to cBOW). The better the token at predicting the surrounding tokens, the higher the quality of the resulting representation

of interest (in a sentence). This approach can be understood as a *moving window* that is constantly attempting to predict its centre (token of interest). This method is several times faster to train than the skip-gram, with slightly better accuracy for the frequent words. An overview of cBOW is shown in Fig. 3.2.

In the **skip-gram** model, the problem is inverted. Each token is used to predict its neighbourhood. This means that a single token input is mapped to a collection of tokens that lie in its surroundings. Skip-gram works well with a small amount of the training data and represents well even rare words or phrases.

The two ideas, albeit similar at first, offer different results and trade-offs when employed in practice. We continue with a more detailed overview of word2vec and related algorithms that were among the first to exploit the two mentioned representation learning ideas.

3.1.2 Method Spotlight: Token Embeddings with Word2vec

The *word2vec* algorithm was one of the first successful implementations of the skip-gram idea discussed in the previous section. This model attempts to maximize word probability occurrence by conditioning on its neighbourhood. More formally, assuming $\{w_1, w_2, w_3, \ldots, w_n\}$ is an *ordered set* of words used as input, the model maximizes the log probability defined as

$$\frac{1}{n} \sum_{t=1}^{n} \sum_{-c \leq j \leq c, j \neq 0} \log p(w_{t+j}|w_t);$$

here, n represents the number of tokens and c the *context size*. The naïve formulation estimates the probability $p(w_{t+k}|w_t)$ via a softmax, i.e.,

$$p(w_o|w_i) = \frac{\exp\left((v'_{w_o})^T v_{w_i}\right)}{\sum_{w=1}^{|W|} \exp\left((v'_w)^T v_{w_i}\right)}.$$

Here, v_w and $v_{w'}$ are the input and output representations of world w, and W is the set of words in the vocabulary. This implementation, however, is impractical due to the large number of words that need to be accounted for each time. To remedy this shortcoming, *word2vec* adopts the idea of *hierarchical softmax*. This improvement overcomes the $\mathcal{O}(|W|)$ evaluation issue that arises by using softmax directly and results in an algorithm that has the complexity of $\mathcal{O}(\log_2 |W|)$, resulting in substantially improved scaling. By representing the output layer with a binary tree ($w \in W$ are leaves) where each node has probabilities of child nodes assigned. This configuration enables each word w to be reached from the root of the tree. If assigning with $n(w, j)$ the j-th node on the path from the root to word w and $L(w)$ the full length of the path, the hierarchical softmax can be defined as

$$p(w|w_I) = \prod_{j=1}^{L(w)-1} \sigma\left([n(w, j+1) = (ch)(n(w, j))] \cdot v_{n(w,j)}'^T v_{w_I}\right),$$

where $ch(n)$ is an arbitrary fixed child of node n and square braces ("[]") denote an operator that evaluates to 1 if the equality holds and -1 otherwise. The sigmoid activation is defined as

$$\sigma(x) = \frac{1}{1 + \exp(-x)}.$$

With this type of probability estimation, the cost of computing the probability of a token given some other token is $\mathcal{O}(L(w_i))$, being on average no greater than $\log_2 |W|$. The approach uses Huffman coding-based construction of the binary tree, resulting in faster lookups for more frequent tokens (and, with it, faster overall training).

The training regime used for *word2vec* adopts additional techniques such as *negative sampling* and subsampling of frequent words. The negative sampling effectively results in the network having to distinguish between the noise distribution and the actual data, increasing the quality of output representations. Substantial improvements to the training regime of *word2vec*-like algorithms are outlined in [60]. Furthermore, extensions for embedding *documents* such as *doc2vec* are also a solid baseline when considering retrieval/classification tasks [73].

3.1.3   Sparse Non-contextual Representation Learning

Even though cBOW or skip-gram-based models remain some of the key methods for obtaining either token or document embeddings, these methods are pre-dated by a collection of approaches that produce symbolic representations – instead of k-dimensional embeddings, sparser, higher dimensional representations are produced. Each dimension in such representation has an exact meaning, for example, word counts and their occurrences. This is also the basis for the next-discussed "Bag-of-words" approach. Texts can be considered as sequential sources of data (token sequences) and, as such, do not directly adhere to learning paradigms that require tabular data (independent dimensions). Hence, a *transformation* is required to obtain a tabular representation from, e.g., a document corpus. A straightforward transformation includes the enumeration of unique tokens and memorization of their counts; this "representation learning" algorithm is called a "Bag of Words." The output of this algorithm is an $\mathbb{N}^n$-dimensional space, where each dimension corresponds to either a single token or an *n-gram* (sequence of n tokens). Values per row represent counts of the text patterns considered. Such representation can be quickly obtained. However, it does not entail any longer range relations (beyond n-grams) and, as such, does not offer as expressive feature space as embedding-based methods. However, a recent work in the domain of natural language processing indicates that such symbolic (explicit) representations can help. Examples of tasks where hybrid representations were considered to achieve state-of-the-art performance include keyphrase detection and document classification.

3.1.4   Method Spotlight: TF-IDF and Beyond

Having discussed one of the simple forms of constructing interpretable (symbolic) representations from documents, we further describe more advanced heuristics that have been shown to offer empirically better results for the tasks of document retrieval, classification, and similarity search-based tasks. Intuitively, the heuristics discussed next attempt to make up for some drawbacks of generic BoW discussed above by incorporating the following ideas:

1. Not all frequent terms are important – frequent at the level of a document subspace can offer different information to generally frequent tokens.
2. Document lengths can be informative of relative token importance.
3. Keyphrases are (unsurprisingly) useful as features, albeit not necessarily directly derivable from raw token counts.
4. Some tokens are not informative in isolation yet play a part in important multi-token interactions.

A prominent example of a Bag-of-Words type of approach that is commonly considered when building lightweight systems is the TF-IDF heuristic. This heuristic is comprised of two terms, *term frequency* and *document frequency*. The term frequency $TF(t,D)$ represents relative frequency of the term t in a document D, i.e.,

$$TF(t, d) = \frac{freq(t, d)}{\sum_{t' \in d} freq(t', d)},$$

where $FREQ(t, D)$ represents the frequency of a term in the document. The second component, the *inverse document frequency*, measures the information content of a word as

$$IDF(t, D) = \log \frac{|D|}{\sum_{d \in D} \mathbb{1}[t \in D]}.$$

From which the final heuristic can be derived as

$$TF\text{-}IDF(t, d, D) = \frac{freq(t, d)}{\sum_{t' \in d} freq(t', d)} \cdot \log \frac{|D|}{\sum_{d \in D} \mathbb{1}[t \in D]}.$$

This heuristic remains useful for text classification in scenarios of noisy data (low-resource settings) [15, 124]. An alternative weighting scheme that is also commonly considered is BM25 [108]. The heuristic is defined as

$$BM25(t, d, D) = IDF(t, D) \cdot \frac{FREQ(t, d) \cdot (k_1 + 1)}{FREQ(t, d) \cdot k_1 \cdot (1 - b + b \cdot \frac{|d|}{avgLen(D)})},$$

where d is the document of interest, D the set of documents of interest, and t the token of interest. The k_1 and b are hyperparameters. This heuristic is used as a complementary approach to neural-only retrieval methods [63]. Using this score as part of the input to neural network-based learners (e.g., BERT) was also reported to improve performance [4] of lexical retrieval. Heuristics such as *TF-IDF* and *BM25* produce *sparse* representations. This implies that any form of efficient learning requires to be adapted to operate with sparse input spaces. Examples of such algorithms include logistic regression and support vector machines [51]. Carefully fine-tuned models that can operate with tens of thousands of sparse dimensions can achieve state-of-the-art performance when considering learning tasks with a small number of samples and high noise levels [83]. We proceed with the discussion of contextual representation learners – neural language models.

3.2   Contextual Representation Learning

Non-contextual representation learning, as discussed in the previous sections, focuses on the notion of "static" embeddings. In recent years, however, the notion of contextual representation learning emerged as the dominant paradigm for many real-life tasks, ranging from translation to document classification and question answering [14, 30]. Compared to non-contextual methods, contextual algorithms are capable of varying the representations based on the context of a token, *dynamically*. For example, the representation of the token "fly" might vary depending on whether one is describing the action of "flying" or the insect. The recent algorithms that adhere to this property are mostly based on the task of *language modelling* and are subsequently termed (large) language models – the focus of the following sections.

3.2.1   Large Language Models

Large language models are a lively research topic that has already seen its success throughout academia and industry. For example, chatbots such as ChatGPT[1] and similar are becoming parts of daily interaction with computers. Similarly, Copilot and similar code augmentation projects[2] enable improved software engineering/development experience. Such applications are possible due to years of research that focused on the task of language modelling. At its core, such endeavours aim to produce systems that, given an existing sequence of tokens, produce the next most probable token [88]. Downstream applications of such models also include representation learning (i.e., production of static embeddings).

We next discuss some of the key breakthroughs in this area that led to existing state-of-the-art and multi-task-solving capabilities.

3.2.2   The Transformer Architecture

One of the undeniable breakthroughs in the area of neural language processing and modelling is the *transformer* architecture [133]. This successor to architectures

---

[1] https://chat.openai.com/

[2] https://github.com/features/copilot

based on, e.g., LSTM and similar recurrent cells adopts a different, simpler paradigm. The transformer architecture follows the *encoder–decoder* paradigm. Here, the input set of tokens is first projected into a latent space (embeddings) and subsequently decoded into an output sequence. The architecture was created with the *auto-regressive* task in mind – outputs of the previous iteration/pass are used as inputs to the next iteration.

The encoder is composed initially of six layers, each comprised of two sub-layers. The first sub-layer is *multi-head attention*, and the second one is a fully connected feed-forward network. Residual connections and layer normalizations are applied as follows:

$$\text{LayerNorm}(x + \text{SubLayer}(x)),$$

where SubLayer(x) is the sub-layer-specific function. The output dimension is pinned to a fixed number (power of 2). The decoder part of the architecture is very similar to the encoder part. Apart from the two sub-layers, however, it also contains the third sub-layer that performs multi-head attention over the output of the encoder stack (i.e., the link between the two stacks). The design also prevents positions from attending to subsequent positions. This masking, combined with the fact that the output embeddings are offset by one position, ensures that the predictions for position i can depend only on the known outputs at positions less than i. The key component of the architecture is the Attention block. In particular, the original transformer uses "Scaled Dot-Product Attention," formalized as

$$\text{Attention}(\mathbf{Q}, \mathbf{K}, \mathbf{V}) = \text{softmax}\left(\frac{\mathbf{Q} \cdot \mathbf{K}^T}{\sqrt{d_k}}\right) \cdot \mathbf{V},$$

where $\sqrt{d_k}$ corresponds to the square root of the embedding dimension, and Q, K, and V represent the "query," "key," and "value" matrices. This mechanism serves as a building block for the multi-head part of the architecture, formalized as

$$\text{MultiHeadAttention}(\mathbf{Q}, \mathbf{K}, \mathbf{V}) = \text{Concat}(h_1, h_2, \ldots, h_n) \cdot \mathbf{W}^O,$$

with

$$h_i = \text{Attention}(\mathbf{Q} \cdot \mathbf{W}_i^Q, \mathbf{K} \cdot \mathbf{W}_i^K, \mathbf{V} \cdot \mathbf{W}_i^V),$$

where, for example, $\mathbf{W}_i^Q$ corresponds to the *projection matrix* – instead of doing the attention directly on, e.g., key embeddings, the values are projected to d_k with a linear layer. These outputs (also for other weight matrices) are used for the final attention computation stated above. Surprisingly, by considering reduced-enough weight matrices (outputs of linear layer), approximately the same amount of compute is required as if considering a single full-dimensional attention layer; multi-head attention serves as a form of *ensemble* of intermediary representations. An overview of the architecture is shown in Fig. 3.3.

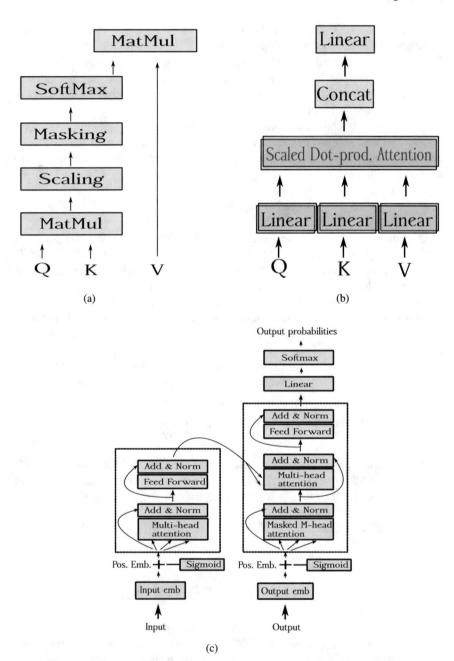

Fig. 3.3 Overview of the attention architecture. (**a**) Scaled dot product. (**b**) Multi-head attention. (**c**) Whole architecture

Transformer-based architectures are to this day the dominant paradigm for sequence-related learning tasks [80] that span from natural language processing to molecular biology (protein folding) [61].

3.2.3 Method Spotlight: Contextual Representations with sentencebert

Of particular focus for this book is the notion of *representation learning*. As with static models, contextual, larger language models are also capable of producing representations. However, as, by design, such models are not built to produce representations but rather solve a particular task (e.g., classification), a research area that focuses on *extraction* of representations from such models emerged. We proceed by discussing, at the time of writing of this book, a state-of-the-art method for obtaining sentence embeddings from BERT-based models, known as the *sentencebert* project [102]. Instead of using BERT-based neural networks end to end (e.g., classification), *sentencebert* explored whether fixed-dimensional representations can be obtained from pre-trained models.

This approach adopts the notion of *siamese* networks for solving tasks that require multi-sentence inputs – an example is the sentence pair classification task (SNLI [11]). The authors of this work also explored different types of aggregations (pooling) to obtain representations of single sentences (embeddings). The considered strategies were utilization of *CLS* token's representations, averaging (*MEAN*) all output representations, and max-over-time (*MAX*) of the output vectors. Empirical results indicate that the *MEAN* approach dominates while being computationally viable for many low-resource projects.

Chapter 4
Graph-Based Methods

4.1 Overview of Graph-Based Learning Methods

We proceed with an overview of the field of graph-based machine learning.

4.2 Selected Methods

The area of machine learning for graph-based data has seen substantial improvements in the last few years. Extensive coverage of all novel approaches is beyond the scope of this book. Hence, the remainder of this chapter will focus on selected use cases to provide the reader with intuition about how some of the well-performing methods work.

4.2.1 Method Spotlight: Representation via Factorization (NetMF)

A canonical task common for graph-based machine learning attempts to address the question of how to map nodes of a graph $G(N, E)$ into a collection of real-valued vectors that can be used as any other embeddings for different down-stream tasks, i.e., $N \rightarrow \mathbb{R}^{|N| \times k}$. Machine learning tasks commonly considered if using node representations directly include node classification, link prediction, community detection, and network pruning. The paper that introduced the NetMF approach was one of the first to *unify* many existing node embedding methods under the same framework – the framework of matrix factorization. Approaches such as DeepWalk [92], LINE [127], PTE [126], and node2vec [46] all offered efficient traversal-based solutions to obtaining real-valued node embeddings ($\mathbf{n} \in \mathbb{R}^k$). We

B. Škrlj, *From Unimodal to Multimodal Machine Learning*, SpringerBriefs in Computer Science, https://doi.org/10.1007/978-3-031-57016-2_4

next discuss the formulations of approaches under a unifying framework, followed by a description of NetMF, a method that enabled improved node embedding based on existing insights. An overview of closed-form matrices representing the process of node embedding for different approaches is stated next. For DeepWalk, the matrix can be summarized as

$$\mathbf{M} = \log \left(\text{vol}(G) \left(\frac{1}{T} \sum_{r=1}^{T} (\mathbf{D}^{-1}\mathbf{A})^r \right) \mathbf{D}^{-1} \right) - \log b.$$

A similar expression was derived for LINE as

$$\mathbf{M} = \log(\text{vol}(G) \cdot \mathbf{D}^{-1}\mathbf{A}\mathbf{D}^{-1}) - \log b,$$

and also PTE, where

$$\alpha_t = \alpha \text{vol}(G_{ww})(\mathbf{D}_{row}^{ww})^{-1}\mathbf{A}_{ww}(\mathbf{D}_{col}^{ww})^{-1},$$

$$\beta_t = \beta \text{vol}(G_{dw})(\mathbf{D}_{row}^{dw})^{-1}\mathbf{A}_{dw}(\mathbf{D}_{col}^{dw})^{-1},$$

and

$$\gamma_t = \beta \text{vol}(G_{lw})(\mathbf{D}_{row}^{lw})^{-1}\mathbf{A}_{lw}(\mathbf{D}_{col}^{lw})^{-1},$$

jointly yielding the final expression

$$\mathbf{M} = \log(\alpha_t; \beta_t; \gamma_t) - \log b.$$

In expressions above, A represents the adjacency matrix, D_{col} is the diagonal matrix with column sum of A, D_{row} is a diagonal matrix with row sum of A, $\text{vol}(G)$ represents the volume of the graph (weight sum), and T and b represent the context size and the number of negative samples. The above expression represents approximations of representations that were shown in existing papers to produce state-of-the-art results. The generalized framework *NetMF* was, based on the insights above, formulated as follows. The version of NetMF for small contexts is shown as Algorithm 1. For bigger context windows, due to computational

---

Algorithm 1 NetMF: small context windows

Compute $\mathbf{p}^1 \dots \mathbf{p}^T$ ▷ Powers of P
$\mathbf{M} = \frac{\text{vol}(G)}{b_T}(\sum_{r=1}^{T} \mathbf{p}^r)\mathbf{D}^{-1}$ ▷ DeepWalk factorization expression
$\mathbf{M}\prime = \max(\mathbf{M}, 1)$
$\mathbf{M}\prime = \mathbf{U}_d \sum_d \mathbf{V}_d^T$ ▷ Singular value decomposition.
Return $\mathbf{U}_d \sqrt{\sum_d}$ ▷ Return final embedding

---

constraints, an approximative extension was developed. Its formulation is seen in Algorithm 2. Substantial speed improvements can be achieved by computing top

---

Algorithm 2 NetMF: large context windows

Compute $\mathbf{D}^{-\frac{1}{2}}\mathbf{A}\mathbf{D}^{-\frac{1}{2}} \approx \mathbf{U}_h \Lambda_\mathbf{h} \mathbf{U}_\mathbf{h}^\mathbf{T}$ ▷ Eigendecomposition (approx)

$\hat{M} = \frac{\text{vol}(G)}{b}\mathbf{D}^{-\frac{1}{2}}\mathbf{U}_h(\frac{1}{T}\sum_{r=1}^{T}\Lambda_\mathbf{h}^\mathbf{r})\mathbf{U}_\mathbf{h}^\mathbf{T}\mathbf{D}^{-\frac{1}{2}})$ ▷ Factorization step

$\hat{M}' = \max \hat{M}, 1$

$\log \hat{M'} = \mathbf{U}_d \sum_d \mathbf{V}_d^T$ ▷ Singular value decomposition.

Return $\mathbf{U_d}\sqrt{\sum_\mathbf{d}}$

---

h eigenpairs via the Arnoldi method. The two formulations demonstrate that many existing methods developed in separate research groups over the past years can be re-iterated as part of the same theoretical framework (with actual implementations). The numerical methods for decomposition are among the most optimized routines in many software packages, making this unification efficient and well performant (as also demonstrated in the paper's results [95]).

4.2.2 Method Spotlight: Learning with Graph Attention Networks

We discussed a unifying approach to *structural* node embedding in the previous section. Even though this branch of methods is highly expressive and efficient in practice, a separate branch of more general algorithms emerged – graph neural networks. One of the first approaches in this area was graph convolutional neural networks (GCNNs) [65]. We next discuss one of the more recent improvements of the idea, Graph Attention Networks (GATs) [135]. This method can be understood as a novel *neural network layer*. For each node, $\mathbf{h}^{|F|}$ vectors are considered alongside the weighted graph as the input (F is the number of features). The layer produces a novel set of features $\mathbf{h}'$ that can have different cardinality (e.g., smaller than $|F|$). Attention coefficients are computed in the next step

$$e_{ij} = \alpha(\mathbf{W}h_i, \mathbf{W}h_j)$$

and can be understood as *importance* of node j's features to node i. In the presented form, all pairwise connections are considered, effectively neglecting any structural information – this is suboptimal. The structure is introduced by *masking*, i.e., e_{ij} is computed only for some neighbourhood of i-th node (including this node). Coefficients are further normalized, so they are comparable. This is achieved via a softmax

$$\alpha_{ij} = \text{Softmax}(e_{ij}).$$

These coefficients are a single-layer neural network in the original paper. Once attention coefficients are obtained, they are further processed via a non-linearity

$$h_i' = \sigma \left(\sum_{j \in \mathcal{N}_i} \alpha_{ij} \mathbf{W} h_j \right) ; \tag{4.1}$$

in practice, *multi-head* attention, similar to the one discussed in the previous section of this book (transformers), is employed, i.e., Eq. 4.1;

$$h_i' = \Big\|_{k \in K} \sigma \left(\sum_{j \in \mathcal{N}_i} \alpha_{ij}^k \mathbf{W}^k \mathbf{h_j} \right) .$$

This process is sensible for all but the final layer of the neural network – in that case, outputs are averaged. The discussed formulation illustrates the key idea of the vibrant field of *graph neural networks* – utilization of graph adjacency matrix as part of the input during neural network training. Even though this branch of methods enables seamless integration of graph adjacency structure into a given neural network training procedure, implementation of actual operations can become one of the bottlenecks when adopting these methods in practice. A comprehensive overview of this branch of methods can be found as [154].

Chapter 5
Computer Vision

5.1   Overview of Image-Based Deep Learning

Image data is inherently *spatial*. Methods that exploit this property were among the first that substantially improved classification performance. One of the first deep learning-based approaches that showed considerable classification improvements on a public leaderboard was AlexNet [70]. This work introduced a handful of algorithmic and implementation improvements. An example is ReLU activation function, defined as

$$\text{ReLU}(x) = \max(x, 0).$$

This seemingly simple activation showed increased noise tolerance and, with it, the performance. It is also computationally feasible when considering multi-GPU training (another contribution of this work). Approaches following this seminal work tend to consider deeper neural network architectures, allowing the approaches to scale even better with the increasing amounts of data. As mentioned in the Overview in Chap. 1, these approaches include Inception V1 [125], VGGNet [121], ResNets [50], ResNeXt [145], and many more [36]. The fundamental building block of current state-of-the-art neural network architectures is *convolutional layers*. This type of layer differs from regular feed-forward (dense) layers by incorporating the local, spatial information by utilizing convolutions as the aggregation mechanism. More recently, however, attention-based architectures started achieving competitive results across many computer vision tasks [47].

5.2   Method Spotlight: Vision Transformers

This section spotlights a recent work on end-to-end object detection with transformers [17]. Object detection refers to identifying a set of *bounding boxes* for objects present in an image. One of the dominant approaches for this task has been YOLO-based convolutional network-based approaches that remain used in many real-life use cases [100]. However, with the advent of attention-based models, attempts to reduce the number of steps and train image segmentation models end to end (i.e., with no intermediary breakpoints that would make them pipelines) became a novel research focus. The discussed DETR approach (DEtection TRansformer) was one of the first such approaches, indicating there exists a link between the type of architectures relevant to computer vision and the ones adopted for natural language modelling.

Conceptually, DETR architecture is composed of two main steps: convolutional network-based feature generation (CNN-based early layers), followed by encoder–decoder architecture (transformer-like approach). The architecture is capable of producing a set of possible bounding box predictions; the approach, during training, utilizes bipartite matching to link predictions to final bounding box definitions. Hungarian algorithm is used to match the ground truth and the prediction [71]. More formally, given a collection of ground truth set of objects y and $\hat{y}$ and a set of predicted objects, the general objective of the architecture is to identify

$$
\underset{\sigma \in \Sigma_N}{\text{argmin}} \sum_{i}^{N} \mathcal{L}_{\text{match}}(y_i, \hat{y}_{\sigma(i)}),
$$

where the loss $\mathcal{L}_{\text{match}}(y_i, \hat{y}_{\sigma(i)})$ is the matching loss between the original object y_i and a prediction $\sigma(i)$. This step produces matched pairs, which are next subject to the following loss function:

$$
\mathcal{L}_{hung}(y, \hat{y}) = \sum_{i=1}^{N} \left[-\log \hat{p}_{\hat{\sigma}(i)} c_i + \mathbb{1}_{(c_i \neq \emptyset)} \mathcal{L}_{box}(b_i, \hat{b}_{\hat{\sigma}}(i)) \right].
$$

The final component that is required for object detection is the loss function that determines the boxes themselves ($\mathcal{L}_{box}$ above). This loss can be defined as

$$
\mathcal{L}_{box}(b_i, \hat{b}_{\sigma(i)}) = \lambda_{iou} \mathcal{L}_{iou}(b_i, \hat{b}_{\sigma(i)}) + \lambda_{L1} ||b_i - \hat{b}_{\sigma(i)}||.
$$

The $\mathcal{L}_{iou}$ [105] is further defined as

$$
\text{GIoU}(A, B) = \frac{|A \cap B|}{|A \cup B|} - \frac{|C \, (A \cup B)|}{|C|},
$$

where A and B are two convex shapes ($A, B \subseteq \mathbb{S} \in \mathbb{R}^n$), and C is the smallest enclosing convex object ($C \subseteq \mathbb{S} \in \mathbb{R}^n$ – note that C is not part of the input but rather computed during the loss itself. We refer the interested reader to [105] for additional implementation details (coordinate-level implementation).

The DETR architecture has shown promising results against strong R-CNN baseline [103], in terms of both speed and average precision. Ablation studies conducted as part of this paper also indicate that both encoder and decoder layers are highly relevant for successful task performance. Omitting layers with regard to the final configuration of 6–6 (encoder–decoder) was proposed as the default strong baseline.

We proceed by discussing selected use cases that illustrate the wide applicability of neural network-based learning for various computer vision tasks. Note that most of the architectures remain largely composed of convolutional and dense layers (or a sequence of such architectures). We discuss two use cases that have seen active research endeavour in the past decade – segmentation of biological images and image generation. We proceed by discussing *image segmentation*, a general task of identifying unique objects in an image.

5.3   Method Spotlight: Biomedical Image Classification with U-Net

Biomedical images are commonly obtained during microscopy, brain scans (EEG and MRI), or similar [1]. They can be two- or higher dimensional; one dimension can also be temporal (e.g., in EEG signals). The next-discussed example focuses on the segmentation of cell images. This task is of high relevance when trying to profile a system's state based on, e.g., ratios of different cell types – neural network-based approaches have been successfully utilized for profiling of not only cancer cells but also bacterial cells in biofilms (layered structures of linked cells). Currently, the gold standard segmentation for in vivo samples remains manual (human) annotation. This process is error-prone, as human annotators can lose focus and are subject to fatigue. The dominant paradigm for unsupervised segmentation adopts deep learning-based methods. Even though useful, these methods remain problematic when amounts of data are too small or domain adaptation, the process of aligning a pre-trained neural network to the new type of data is an issue. A typical CNN architecture consists of many convolutional layers (hidden, input, and output). Each layer has fully weighted connected neurons/nodes that seek definite attributes from the prior layer feature map based on the kernel size that was used within that specific layer. The feature map dimensions continue to increase as we delve deeper because additional information is indexed by the networks. However, the dimensionality can be downsized by pooling layers to lower network complexity. According to recent publications, incorporating a number of small-sized filters (typically 3×3) can be useful in obtaining a desirable performance.

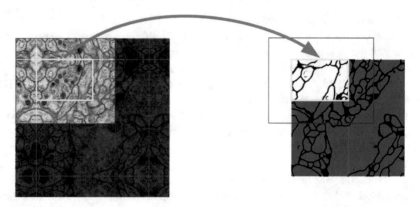

Fig. 5.1 An example of cell segmentation [109]. The blue area represents the space of the image where segmentation takes place, and the yellow area is the focus area. The output is segments – parts of the image that represent distinct objects; in the case of this image, these are cells

This is despite the fact that the pooling layer typically has a filter size of 2 and stride=2. Zahangir [6] incorporated U-Net architectures (R2U-Net and RU-Net) in a Recurrent Convolutional Neural Network (R-CNN) for various applications in the field of medical imaging, including lung cancer segmentation, retina blood vessel segmentation, and skin cancer segmentation. We will next discuss U-Net architecture, one of the flagship "backbone" architectures commonly used in this domain [109]. An example of the task solved by the seminal U-Net paper is shown in Fig. 5.1. The architecture of U-Net is summarized in Fig. 5.2. U-Net architecture is among the most efficient architectures for image segmentation. Its success can be attributed to both the architecture and computational efficiency. However, the original work also emphasizes data augmentation as the process that enabled the detection of invariances (shift and rotation). One of the key components of U-Net is the computation of separation borders via morphological operations. In particular, the weight map, a structure that enables a better account for pixel imbalance throughout the training data, is computed as

$$\omega(\mathbf{x}) = \omega_{\mathbf{c}}(\mathbf{x}) + \omega_{\mathbf{0}} \cdot \exp\left(-\frac{(\mathbf{d_1}(\mathbf{x}) + \mathbf{d_2}(\mathbf{x}))^2)}{2 \cdot \sigma^2}\right),$$

where ω_c is the weight map that balances class frequencies, d_1 denotes the distance to the border of the nearest cell, and d_2 is the distance to the border of the second nearest cell. ω_0 and σ are free parameters (pixel sizes). Gaussian-based weight initialization is used to ensure stable network initialization. The discussed branch of approaches is most effective when the data sets they use in the training phase are large enough. However, the number of images available for training is limited for some organs. These methods can also be slow, although the parallel process on the network layers and the use of modern GPU cards can mitigate this drawback.

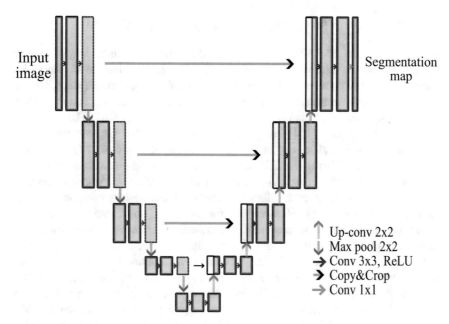

Fig. 5.2 U-Net architecture [109]. The architecture enables incremental compression of the input image and, with it, different granularity of approximation

They are suitable for almost all image modalities. However, training data sets are not always freely available to the public because of privacy issues related to patients.

5.4   Method Spotlight: Generative Adversarial Networks

Apart from the detection of specific objects and the classification of images into pre-defined class categories, the recent wave of neural network-based methods also offered substantial improvements in the area of *generative* modelling. Here, the main goal is to be able to generate an image based on either text-based description (which is already a multimodal endeavour) or simpler, generating images with similar styles by first training models to approximate an image and subsequently vary the latent space's properties so similar, albeit different outputs are obtained. The idea discussed in this chapter is that of *generative adversarial* (neural) networks (GANs) [42, 43]. The idea of GANs revolves around two types of neural network architectures: *discriminator* and *generator* architectures. Their tasks differ, yet when combined they enable high-quality image generation.

The generator (G) and discriminator (D) networks are trained in a two-player minimax game that can be summarized as

$$\min_G \max_D = \mathbb{E}_{\mathbf{x} \approx p_{\text{data}}(\mathbf{x})} \big[\log(D(\mathbf{x})) \big] + \mathbb{E}_{\mathbf{z} \approx \mathbf{p}_z(\mathbf{z})} \big[\log(\mathbf{1} - \mathbf{D}(\mathbf{G}(\mathbf{z}))) \big].$$

Theoretical results of [42] indicate that in the parametric limit, data generating distribution can be retrieved. In practice, K and D are optimized in an alternating manner – discriminator network is commonly optimized for multiple steps, with an intermediary optimization step that concerns the generator network. Training GANs is notoriously difficult, as the convergence of both networks needs to be preserved. In a follow-up paper, [114] introduced multiple ideas that improve the general performance/training efficiency of GANs. For example, *minibatch discrimination*, the process of considering multiple samples of data (images) simultaneously rather than in isolation, offered stability improvements. Further, *historical averaging*, the process of including past values of parameters during loss computation, also improved convergence. *Label smoothing*, the process of transforming a label space from a discrete into a continuous one that was shown to improve resistance to adversarial examples. The final idea that substantially contributed to performance was *virtual batch normalization* – instead of considering common batch normalization that can suffer from co-dependence between different inputs, normalizing batches with regard to *reference batch*, a batch of data selected prior to training. This step, albeit useful for denoising and convergence improvements, yields a computationally more expensive optimization (two forward passes), so it is not considered for the discriminator part of the common GAN systems.

Part III
Multimodal Machine Learning

Multimodal machine learning is an exciting and rapidly evolving field that has gained significant attention in recent years. The ability to process and learn from data across multiple modalities has opened up new possibilities in a wide range of applications, including speech recognition, natural language processing, and image recognition.

The use of multiple modalities in machine learning has become increasingly important due to the fact that many real-world applications require the integration of information from multiple sources. For example, in autonomous driving, it is necessary to process information from cameras, lidar, and radar sensors to make decisions in real time. In speech recognition, audio and text modalities are often combined to improve accuracy.

One of the key challenges of multimodal machine learning is how to effectively combine information from different modalities. There are different approaches to solving this challenge, including fusion-based approaches, graph-based approaches, and attention-based approaches. Fusion-based approaches involve combining multiple modalities at the input level, feature level, or decision level. Attention-based approaches use attention mechanisms to learn to focus on relevant parts of the input from different modalities. Another important aspect of multimodal machine learning is how to evaluate the performance of models. Traditional evaluation metrics, such as accuracy and precision, are often insufficient in the context of multimodal machine learning, as they do not take into account the interactions between different modalities. Therefore, researchers have developed new evaluation metrics, such as cross-modal retrieval and cross-modal classification, that take into account the interactions between different modalities.

In summary, multimodal machine learning is a rapidly evolving field that has the potential to revolutionize many areas of science and industry. By effectively combining information from different modalities, researchers can develop powerful machine learning algorithms that can handle complex real-world scenarios. However, there are still many challenges to overcome, including how to effectively combine information from different modalities and how to evaluate the performance of models.

Chapter 6
Multimodal Learning

6.1 Overview of Multimodal Machine Learning

Multimodal machine learning has been considered alongside other branches of machine learning and AI in the past 50 years. For example, the seminal paper [85] already investigated the origins of speech recognition from visual and audio signals. This work had a considerable impact on many methods subsequently developed.

The domain of human–computer interaction became one of the key frontiers for a better understanding of interfaces for communicating with a machine [58]. For humans, speech- and text-based inputs are very common (and simple to capture). Hence, these two modalities were among the well-studied ones in the earlier days. After the 2000s, many open challenges were proposed that considerably improved the methodology in the field. An example is Avec 2011 [117] audio-visual emotion challenge.

More recent work also focused on video summarization [34], a modality previously rarely considered due to computational constraints. Multimodal inputs are also suitable for robotics-related problems [66]. A recent overview exposes key open challenges of multimodal learning that represent the common thread between many sub-domains of multimodal learning [77]. The first domain of interest is multimodal alignment; this concept attempts to establish the laws and general properties that govern interaction between different modalities. Of particular interest in recent years remains cross-modal interaction identification – even if independent modalities do not contain the information required to solve the task of interest, their interaction might be sufficient. The second property actively studied is that of *compositionality*. It has been consistently shown that neural network-based approaches adhere to some form of compositional learning – the concepts are distilled hierarchically, from raw, signal-level information to more coarse-grained concepts in later layers/within the network. Understanding how different types of representations combine and interact to form more meaningful representations that enable *reasoning* remains a challenging task. This endeavour frequently attempts to exploit *external* knowledge

© The Author(s), under exclusive license to Springer Nature Switzerland AG 2024
B. Škrlj, *From Unimodal to Multimodal Machine Learning*, SpringerBriefs in
Computer Science, https://doi.org/10.1007/978-3-031-57016-2_6

to facilitate out-of-distribution learning. Of particular interest in recent years is also the notion of representation generation. Here, the key focus is the study of how to generate representation/modality of interest effectively and adequately. The two tasks commonly considered are information reduction (summarization) – this property is useful, for example, for retrieval-based tasks and modality translation – the process of translating one modality to another.

The recent encoder–decoder-based architectures have shown to be a promising avenue for exploring this type of task. The final task that is commonly considered, primarily when focusing on larger models, is that of *transferability* of representations. Nowadays, vision and text models are seldom trained from scratch. By being able to jump-start a learning process by using a *pre-trained* model for a given modality of interest, efficient learning can take place without the need for specialized hardware. Further, transferring information can be considered either within the same modality (e.g., transfer learning for document representations), as well as *across* modalities – this problem, in particular, has been of recent interest, as text and video-based inputs are increasingly more frequent, and enable concise studies of transferability across the two modalities [20].

A vibrant research area in recent years has also been that of *representation fusion*. This subfield of multimodal machine learning attempts to identify general patterns that govern the combination of different modalities. As modalities and data of interest are necessarily at different semantic levels, fusing representations can be challenging. For example, fusing image and document-based representations is seemingly straight-forward (assuming no specialized normalization is required); however, if input data is also that of, e.g., concepts and relations between them (i.e., a knowledge graph-like data), such information needs to be first embedded and only subsequently merged/combined with other representations. Such *fusion* of representations is one of the key topics of multimodal learning and is present in the majority of recent work that focuses on the efficient and scalable inclusion of multiple modalities.

6.2   Fusion Algorithms

The two distinct types of fusion in multimodal machine learning are early fusion and late fusion [8]. Early fusion involves combining the input data from multiple modalities at the very beginning of the learning process, resulting in a joint representation of the data. This joint representation is then used for further processing and prediction. On the other hand, late fusion involves processing each modality separately and then combining the predictions at the end of the learning process. Late fusion can be further divided into two subtypes: feature-level fusion and decision-level fusion. Feature-level fusion involves combining the features extracted from each modality before making predictions, while decision-level fusion involves combining the predictions made by each modality after they have been processed separately. Each type of fusion has its own advantages and disadvantages, depending on the specific application and the nature of the data being used. It is important to carefully consider these factors when choosing a fusion strategy for a multimodal machine learning system.

6.2.1 Early Fusion

In early fusion, the input data from different modalities is combined before any processing or learning takes place. This can lead to a joint representation of the data that can be used for further processing and prediction. Studies have shown that early fusion approaches are particularly effective for multimodal convolutional neural networks. In fact, a recent study demonstrated that early fusion methods outperformed late fusion methods for the task of emotion recognition using audio and visual modalities [38]. Another advantage of early fusion is that it can help to reduce overfitting, as the model is forced to learn a joint representation of the data from the different modalities. However, early fusion approaches can also be more computationally expensive and may require more memory than late fusion approaches, as the joint representation can be larger than the individual representations of each modality.

Early fusion has been successfully applied in a wide range of multimodal machine learning tasks. For example, in the field of computer vision, early fusion has been used for tasks such as object recognition, image classification, and facial expression recognition [153]. Similarly, in the field of natural language processing, early fusion has been used for tasks such as sentiment analysis, text classification, and machine translation [39]. Early fusion has also been used in the field of speech recognition, where it has been shown to improve the accuracy of speech recognition systems by combining information from both audio and visual modalities [93]. Furthermore, early fusion has been used in biomedical applications, such as in the analysis of medical images and the prediction of disease outcomes based on clinical data [123]. These examples demonstrate the versatility and effectiveness of early fusion in a wide range of multimodal machine learning tasks. A schematic overview of early fusion can be seen in Fig. 6.1.

6.2.2 Late Fusion

Contrary to early fusion, late fusion corresponds to the process where modalities are initially processed separately (e.g., different feature extraction/embedding layers); however, at the intermediary stage of learning, they are joined as part of the same parameter space that is jointly considered for subsequent learning. In practice, this type of fusion is mostly synonymous with deep learning-based learning. This is due to the modularity of contemporary architectures, where modalities can be

Fused representation

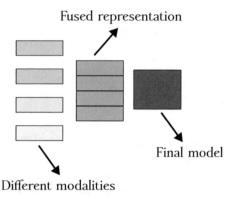

Final model

Different modalities

Fig. 6.1 Early fusion. The process of early fusion involves an early step, most commonly at the representation layer (feature space) that creates the joint representation. This representation serves as input to the final learner. Fusing modalities can be done via intermediary models and is considered early fusion as long as intermediary embeddings are fused prior to being fed to a separate model

initially processed with separate blocks but joined into the final, e.g., classification head. Such approaches are, for example, adopted when designing architectures for autonomous driving (different sensory inputs). An example of late fusion is that of majority voting [91], a strategy where different modalities provide the class labels, and the most frequent one is used as the final prediction – note the analogy to ensemble-based learning (e.g., random forests) [13].

Late fusion has been widely used in various applications, such as multimodal sentiment analysis, activity recognition, and object recognition. For example, in multimodal sentiment analysis, late fusion can be used to combine the predictions made by models trained on textual and visual data to improve the accuracy of sentiment classification [151]. Similarly, in activity recognition, late fusion can be used to combine the predictions made by models trained on accelerometer data and audio data to improve the accuracy of activity recognition in real-world scenarios [41]. Finally, in object recognition, late fusion can be used to combine the predictions made by models trained on different image scales to improve the accuracy of object detection [38]. Overall, late fusion is a powerful technique that can be used in a variety of applications to improve the accuracy and robustness of multimodal models. A schematic overview of late fusion can be seen in Fig. 6.2.

6.2.3 Learning from Images and Text

We proceed by describing selected approaches that jointly consider image- and text-based data. These two modalities are among the most common ones, explaining why existing research endeavour often focuses on the tasks of, e.g., multimodal classi-

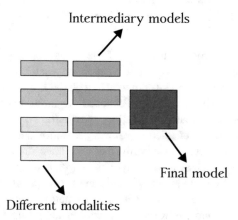

Intermediary models

Final model

Different modalities

Fig. 6.2 Late fusion. The process of late fusion differs from early fusion in the early step – instead of creating a joint feature space, late fusion commonly considers intermediary *models* that normally output one or more predictions. The combination of outputs forms different representations at the model output level is done via a separate model. One of the key properties of late fusion is its modularity – interchanging models and testing their separate impact is simpler compared to working directly at the representation level

fication or regression. As discussed in Chap. 2, images and text require dedicated approaches to encode the information in either high-vocabulary token sequences (text) or more spatial, higher dimensional objects (image multidimensional arrays). Many state-of-the-art approaches build on top of existing approaches that primarily focus on specific modalities yet can be combined to serve as multimodal classifiers.

This section follows the concepts in a recently introduced *taxonomy* of multimodal machine learning [7]. Joint learning from images and text is one of the most ubiquitous forms of multimodal learning. Methods capable of such learning include neural networks, graphical models, and other sequential methods. On a higher level, tasks that concern images can be split into multiple possible categories. The first practically useful example is that of *retrieval* – the process of specifying examples to the system with the goal of retrieving similar samples. Practical applications of this idea include image captioning [35], media retrieval [147] (search), and visual speech [12].

A separate branch of methods is of more *generative* nature. Here, image and video description generation are two everyday use cases [76, 129]. The main objective of systems aimed at description generation is the *fast* and accurate generation of shorter text segments that can be annotated. The third branch of methods adheres to the encoder–decoder principle. Here, inputs of one modality are effectively *translated* to a different modality. Example use cases include image captioning [68], video description generation [150], and text to image synthesis [101].

6.2.4   Learning from Sound and Text

Audio-visual speech recognition is one of the oldest forms of multimodal learning [18]. The first systems for text to speech originated already in the 1950s. However, the first more general system focusing on the English language was developed in 1968 [69]. Early systems first exposed the input text to linguistics-based approaches that distilled phasing, intonation, and duration of words. Translating this information into phonemes enabled waveform generation and effective speech [33].

More recently, however, neural-only methods have dominated this field. An example is the FastSpeech system [104]. This work, as many others in the domain of contemporary multimodal learning, builds on the *transformer*-based architecture. The work addresses some of the limitations of neural-only approaches that are absent for statistical (parametric) approaches considered up to recent years; the problems include word skipping and voice control issues. FastSpeech adopts the feed-forward transformer architecture, where phoneme embeddings are sent through a length regulator operator and decoded into the final waveform. Length regulator performs on-the-fly duration prediction, which helps overcome one of the issues of neural-only approaches – inconsistent phoneme output. This work also demonstrates that a teacher–student paradigm is sensible for this type of task. Attention alignments are extracted from the trained (teacher) model.

The attention block with the highest "focus" is selected as the basis for phoneme duration sequence calculation. The approach was one of the first to address the end-to-end notion of text-to-speech training with transformer-based architecture. This work also demonstrates a clever use of attention heads/selection that enables better convergence/focus on the task at hand – one of the main results is the fact that only certain attention heads are in the final stage of training relevant to the task at hand (others can contain noise that is best pruned). Work like FastSpeech demonstrates the current trend of multimodal text-to-speech training; by considering a single architecture (or a teacher–student-like scheme), this type of approach enables more scalable and transferable learning.

6.3   Use Cases of Multimodal Learning

In this final chapter, we discuss some of the existing use cases of the multimodal learning paradigm on real-life data. The purpose of this chapter is to highlight which existing data sources already adhere to the multimodal paradigm and are, as such, readily available for future endeavours in this area. Furthermore, we also discuss some of the drawbacks of the mentioned methods with the aim of pinpointing interesting research questions that arise when considering multimodal inputs (and outputs).

6.3.1   Classification of Recipes and Food-Related Data

A rich source of data is that of recipe information. This open data is primarily built for humans and is often characterized by rich text descriptions and image-based supplementary material. Recipes need to be verbose and concrete for them to be successful, making this type of data a good use case for testing the capabilities of multimodal machine learning.

For example, [149] investigated how CNN-based features can be combined with word2vec-based text representation and have shown that considerable classification improvements can be achieved if both modalities are considered. The task addressed was taste prediction. Their work also showcased late fusion as the dominant fusion type, achieving substantially better results than early fusion.

Recipes have also recently emerged as an exciting source of data for the evaluation of question-answering systems (QASs) [148]. For example, RecipeQA is a data set comprised of approximately 20K instructional recipes with multiple modalities such as titles, descriptions, and aligned sets of images. With over 36K automatically generated question–answer pairs, we design a set of comprehension and reasoning tasks that require a joint understanding of images and text, capturing the temporal flow of events and making sense of procedural knowledge. Another prominent recent data set is the UPMC-Food-101 data set [138]. This data set consists of about 100,000 recipes for 101 food categories. A recent paper indicates that the BERT-based model, combined with the CNN-based vision model, achieves state-of-the-art performance on this data set [40]. Their approach indicates that early fusion is a viable alternative for noisy and bigger data sets such as UPMC-Food-101. An early fusion approach termed MuRE (Multimodal Representation Evolution) that was subject to evolution-based meta-learning, i.e., the process of automated tuning of obtained representations for a given task, has also shown a strong performance on multimodal problems that include recipe-based data [122]. More recently, transformer-based models have emerged as the branch of methods that offer competitive performance when classifying recipes (in a multimodal setting) [81].

Apart from classification-based tasks, multimodal aspects of recipe-based data were also considered in the context of *information discovery*. For example, [89] proposed a multimodal deep belief network, a type of neural network-based approach that yielded representations directly useful for tasks of augmented image retrieval and ingredient inference. Both tasks serve as the building blocks for constructing systems that enable users to find and discover new information.

6.3.2   Multimodal Retrieval

The topic of multimodal retrieval has seen wide adoption throughout academia and industry, powering solutions ranging from fashion sites [78] to multimodal search [22, 79]. An example framework that enables retrieval across modalities such as audio, video, and text has been proposed by [99]. This work computes a joint representation from a collection of unimodal ones. The representation is automatically re-weighted to accommodate for "alignment" between the constituents. The method can also handle new data by embedding it into the existing space, making it computationally efficient. Methods like this have become the common paradigm for approaching multimodal retrieval. More broadly, early or late fusion is considered to compute the similarity score or representation. Next, the obtained score is used for efficient retrieval without re-computing representations too frequently.

The paradigm of multimodal retrieval has also been considered in a biomedical context [16]. This methodology pre-computes a collection of "fingerprints" for images and jointly with a latent semantic analysis-like approach – an approach to unsupervised topic detection – computer codes for retrieval. An additional focus of this work is *missing modalities*, the issue common to real-life data; even though there are many multimodal, e.g., image–text pairs, it is common that either image or text-related information is missing. Building a system that robustly handles this aspect is essential in practical scenarios. This issue was addressed by training separate deep belief networks (DBNs) that are robust to missing data – the late fusion-like approach ensures that information, albeit incomplete, gets to the final parts of the network and can be fully exploited.

6.4   Multimodal Language Models

Since the creation of the transformer architecture [133], language models have consistently shown promise when modelling human (and other) languages. Recently, however, the paradigm was successfully expanded to the domain of visual and sound-based understanding. Current state-of-the-art language models such as GPT4 (December 2023, online API) enable users to process text and image-based prompts. At the time of writing this section, the dominant language models remain closed-sourced, even though technical reports on their predecessors are available and will be discussed in this section. Open-source models such as LLaMA [130] and Alpaca [128] are showing promise, even on commodity hardware. However, currently, they remain a separate, lower performing class of models compared to the commercial alternatives.

The multimodal aspect of language models is achieved by combining transformer-based architectures with vision models, which are increasingly also transformer-based. However, convolutional neural network layers can also play a role in segmentation/input preparation. Models that are based around ideas of language, however, consider different modalities were already considered by [67]. The authors have shown that text features can be obtained jointly with image-based features (from convolutional neural networks). With the advent of language models, amounts of data to account for different modalities started to become increasingly smaller. For example, [131] investigated the idea of multimodal few-shot learning with "frozen" language models; weights of the language models were not updated, and by prompting, multimodal performance/capabilities could be achieved.

One of the recent state-of-the-art multimodal language models is PaLM-E [31]. This model *transfers* knowledge from visual language domains to a joint, embedding-based reasoning regime where it can generalize and reason about existing and new concepts that are presented to it. One of the key ideas of this paper is the notion of *multimodal sentences*. Apart from considering token sequences as is commonly done by language models, this model considers images, 3D representations (models), and states of an object. By creating such "joint" input space, scalable training and fine-tuning can be achieved, which is demonstrated by the work. The authors also observed that scaling the language model increases its performance and memory retention on language-only tasks – problems where other multimodal models tend to lose performance.

Benchmark data sets for multimodal learning experiments have also advanced significantly in the past few years. For example, the MME benchmark [37] consists of 14 tasks that measure both perception and cognition-related capabilities. Tasks related to perception include:

1. Existence (presence detection). This task focuses on identifying whether an object is present in a given frame of reference.
2. Count of objects. Given a pivot/seed object, the task entails counting appearances (with minor variations) of the object within an image.
3. Position of objects (e.g., where is the object in the image). Positions can be specified relative to some reference frame or as absolute coordinates.
4. Colour of objects of interest. Colour detection attempts to distil whether parts of the spectrum were not captured during training, as this might lead to adversarial examples.

Fine-grained perception tasks include:

1. Asking for "Poster," i.e., the author of an object (e.g., an image or a movie). This task aims to probe the model for understanding the links between different types of objects and reference people. Objects can be drawings or other pieces of art.
2. Celebrity detection. Celebrities are commonly present throughout the text-based training data (for language models). This task is effectively testing retrieval capabilities for concepts that are known to have good coverage and are frequent in training data.
3. Scene description. Identifying objects and capturing relations between them is not a trivial task. Models capable of scene description understand the object that composes a scene.
4. Landmark detection. Landmarks are, similarly to the celebrity task, known parts of nature (man-made or not) that are present in training data. Landmarks can be part of noisy inputs and hence present an interesting learning endeavour.
5. Artwork identification. Similar to "Poster" identification, retrieval of properties about a piece of art involves training models on large collections of artwork data.

The precise perception task also includes phone number identification from images. On the other hand, cognition-related tasks include:

1. Commonsense reasoning. Commonsense reasoning tasks are at the forefront of tests for understanding the general capability of multimodal models. They include simple causal relationship induction (e.g., if an apple fell from the tree, where is it now?) to more involved tasks that involve multi-hop reasoning.
2. Numerical calculation. Calculation has been a non-trivial task for language model-based approaches. It entails models understanding types of numbers and relations between them that generalize.
3. Text translation. Text translation is one of the most useful tasks that, at the same time, tests models' capability to map out objects in two domains (languages) correctly. The details during translation are subtle and pinpoint corner case behaviour of existing models.
4. Code reasoning. Aking models to understand source code and reason about it is one of the most useful tasks for software engineers and other developers. By reasoning about a piece of code, models are tested to predict outputs and other corner case behaviour correctly.

The considered tasks are all solved by the same modal, which indicates its general learning and reasoning capability. With the rise of data sets like MME, more involved evaluation schemes can be tested, learning to a more precise understanding of the general intelligence/capabilities of multimodal language models.

6.5   Method Spotlight: Mixture of Experts (MoE) Architecture

We conclude this section by highlighting an ensemble method that has offered consistently good language modelling performance when considering smaller language models (models that consist of, at most, tens of billions of parameters). The MoE architecture is depicted in Fig. 6.3. The MoE approach, conceptualized in the early 1990s [57], attempts to *branch out* parts of the computation to multiple computationally simpler units (when compared to more complex single-block-like models). Contemporary MoEs can be pre-trained with fewer resources, implying better scaling regarding the number of tokens (data) or the parameter count. One key insights of MoEs are *sparse* layers [107]. These replacements to commonly used dense (FFN) layers are computationally more feasible, as they entail only parts of the parameter space [32]. The "experts" in a typical MoE architecture are FFNs, as any other, with the main difference being that they are commonly smaller. More recent attempts also tend to explore hierarchical expert schemes.

The second key component of MoE is *routers*. In the context of LLMs, this component is responsible for sending which tokens to which expert. Routers determine the distribution of inputs for the experts' space. There is no constraint

Fig. 6.3 The MoE
layer [119]. Two main
components of the MoE layer
are the *router* network
(yellow) – this component
decides, on the fly, which
experts, the second relevant
component, to utilize for a
given input

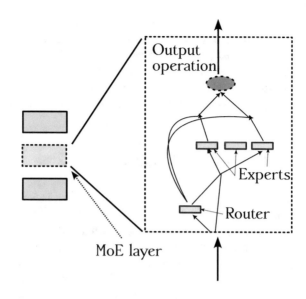

that would specify how many tokens need to be sent per expert (not necessarily a
single token), even though this has repercussions for the computational complexity
of the MoE. The router is pre-trained alongside MoE – this is a learnable component
that is aligned and tuned alongside the operation of downstream experts in a given
MoE. One of the issues with training an MoE model is the trend to "overfit" to only a
few experts while leaving others intact. This issue can be overcome by manipulating
the loss function so that other experts' inputs are also accounted for more frequently.

One of the main benefits of MoE that make this paradigm attractive for live
systems such as ChatGPT or similar is its implications for inference. Even though
the MoE, as such, can have a similar number of parameters to larger models,
only specific subspace(s) is used during inference, drastically reducing inference
time. The remaining issue is memory utilization; all experts need to be loaded in
RAM, which makes this type of approach memory-intensive. Memory mapping is
a possible alternative, even though for this solution, the impact on inference can be
hard to predict. Note that when storing an MoE, there are two main components: the
shared core part of the architecture and a collection of independent experts. Models
share the core part of the architecture, making the memory footprint potentially
smaller. In general, MoE is suitable for multi-machine deployments, as computing
can be directly split across experts and combined when relevant.

6.6 Future Prospects

In the final part of this book we overview the discussed topics and inspect potential
directions of existing research based on the fast-moving trends of 2020s. Through-

out this book, an overview of the origins of machine learning that led to current, multimodal multi-task-capable systems was offered. The rate of progress has been unprecedented in the recent years and remains surprising to many practitioners in the field. We continue the discussion with a collection of topics that are slowly emerging as relevant (albeit some were previously considered before, but the time not appropriate due to either compute or data-related issues).

6.6.1   Video and 3D Models

In recent years, 3D model-based inputs have become increasingly relevant in the field of multimodal learning. While conceptually similar to image-based inputs, 3D models offer a synthetic representation of an object that can provide much more fine-grained information about its properties and relationships. This can be particularly useful in applications where precise understanding of object shape and structure is required, such as in medical imaging or robotics. By using 3D models as input, multimodal models can better ground core concepts related to abstract shapes and sub-shapes that appear in actual images, leading to improved performance and robustness.

In addition to 3D models, video-based data is also of increasing interest in the field of multimodal learning. While similar in the concept to image-based data, video data comes with a unique set of metadata that is commonly attached, such as subtitles and captions. This metadata provides a rich training ground for exploring the capabilities of multimodal models, allowing them to learn from both visual and textual information. However, the computational complexity of real-time video understanding remains a significant bottleneck, limiting the wider adoption of multimodal models in this domain. Nevertheless, solving this problem will enable new approaches for dynamic, time-dependent data to be considered on commodity hardware, opening up new possibilities for applications such as video analysis, surveillance, and autonomous driving. Furthermore, combining video data with audio and text can improve speech recognition accuracy. Another example is in video surveillance, where multimodal models can be trained to detect abnormal behaviour in crowds by analysing both visual and audio cues [56]. In the field of autonomous driving, multimodal models can be trained on video data from various sensors such as cameras and lidar to improve object detection and tracking accuracy [144]. However, the computational complexity of real-time video understanding remains a significant bottleneck, limiting the wider adoption of multimodal models in this domain. Nevertheless, solving this problem will enable new approaches for dynamic, time-dependent data to be considered on commodity hardware, opening up new possibilities for applications such as video analysis, surveillance, and autonomous driving.

6.6.2 Time Series-Based Data

Time series-based data is a prominent type of data that is ubiquitous across many fields, from finance to electrical engineering to biological clocks. While the time dimension can be part of image-based data, patterns in the time series themselves can be used to improve inference and estimate the overall trend of the target variable. Historically, models that have shown good performance for one-dimensional time series data were mostly based on recurrent neural network-based approaches or pattern mining-based approaches. For example, Long Short-Term Memory (LSTM) networks have been widely used for time series prediction due to their ability to capture long-term dependencies in the data.

Recently, however, transformer-based approaches have started to show promising results in this domain. For example, Wen et al. [141] proposed a transformer-based approach that outperformed traditional LSTM-based approaches on several time series prediction tasks. Similarly, Masini et al. [84] proposed a transformer-based approach for time series forecasting that also achieved state-of-the-art performance on several benchmark data sets. Overall, transformer-based approaches offer a promising alternative to traditional recurrent neural network-based and pattern mining-based approaches for time series analysis and are an active area of research in the field.

6.6.3 Multimodal Algorithmic Paradigms

Throughout the book, many different approaches were discussed. Two main paradigms to multimodal learning can be described as follows:

Multimodal representations as inputs, output of single modality. This branch of methods focuses on constructing either ensembles of models, representations, or a single model that can take up different modalities as inputs. Here, the goal is to identify the appropriate constellation of neural network layers, weights per representation, or appropriate models for individual modalities.

Multimodal learning, end to end. Here, one or more modalities are considered as input, and one or more modalities are considered as output – the modalities considered are not necessarily the same (e.g., text to speech). This branch of methods are mostly neural network-based, as these models enable seamless modality fusion within the model itself.

The algorithmic aspects that are an active research area differ for the two types of multimodal systems. On the one hand, multimodal systems that map to a single modality (target space, in this case) often suffer from expensive training that requires custom hardware (e.g., General Processing Units – GPUs). By being able to speed up training, approaches can become much more useful in low-resource scenarios where sufficient infrastructure is not necessarily available. Furthermore, if each inference step requires pre-computing representations via multiple separate models, this can incur additional time cost that can be prohibitive in practice. A promising paradigm to mitigate some of these issues is *model pre-training*. Models, if enough high-quality data is available, can be pre-trained and subsequently only fine-tuned for a designated downstream task at hand. Expensive pre-training can be done on distributed hardware architectures (clusters); however, fine-tuning of such models can be done on commodity hardware (e.g., laptops). If adopting a multi-representation approach, i.e., computing multiple representations that are used as input for a learning algorithm, representation *quantization* can also be considered. This relatively young field arose due to the need for faster inference when considering larger models (billions of parameters). As soon as neural network-based representations are considered, quantized representations are a viable option to achieving faster inference time (and representation storage benefits).

The second branch of algorithms that consider different modalities within the same neural network architecture can be optimized with similar approaches. Model quantization, as well as teacher–student network-based approaches, can offer more compact models that scale substantially better.

We finally discuss the resource consumption of methods and systems discussed in this work. In general, resource consumption of a given system is highly dependent on the type of modality considered. For example, at the time of writing of this book, large language models dominate in terms of size. By being able to utilize such models even for single modality tasks, considerable resources are required; inference is possible without specialized hardware; however training is for now limited to entities in possession of large GPU compute clusters (or similar hardware). On the other hand, fine-tuning language models, as well as computer vision models, is within reach for an average workstation-level hardware and is as such more widely available. Other types of modalities such as graph-based data or time series might require much lower footprint to obtain adequate representations that can be used for multimodal "early fusion" type of approaches. Interestingly enough, complementarity of representations can outweigh a single modality, large neural network – if information available to the system is complementary, utilizing all aspects, albeit more poorly, can be net positive.

6.7   Conclusion

This book offered an overview of machine learning from the point of view of different modalities considered. Beginning with unimodal learning that is nowadays in widespread use, we proceeded by discussing multimodal approaches that attempt to leverage different aspects of the information available. In particular, this book focuses on the most common modalities such as texts, images, and graph-based data, even though modalities such as time series and similar signals are also considered.

The first part of the book (Chaps. 1 and 2) focuses on background on machine learning, historic overview, and some of the key techniques that kicked off the current wave of embedding-based machine learning methods. In particular, we discuss the two paradigms for learning from text-based data: the embedding-based techniques and symbolic techniques. Both paradigms are equipped with examples that represent the main ideas of either domain, from the initial word2vec (and extensions) to contextual language models that are becoming the foundational technology for dealing with many other modalities. We discussed construction of TF-IDF-based symbolic spaces and what such representations entail, followed by construction of contextual representations. The graph-based methods highlighted in this book were selected to illustrate one of the key ideas in graph deep learning – graph neural networks (graph attention networks, in particular). Further, we also discussed NetMF, a methodology that unified many other paradigms for structural node embedding (nodes have no features).

We concluded the overview of unimodal methods by describing some of the key ideas from computer vision. By focusing on image segmentation (U-Net) and vision transformers, we illustrate the current trend of end-to-end, transformer-based vision algorithms. We concluded the chapter by discussing generative adversarial networks (GANs) that kicked off a wave of progress in generative machine learning (focusing on images).

In Chap. 3 we focused on multimodal machine learning. This chapter attempts to entail some of the fast-growing fields, where each month, substantial achievements are reported. The purpose of the chapter is to demonstrate how existing multimodal approaches and techniques can already provide cross-context, cross-modal learning. This type of learning is paramount to solving certain tasks where a single modality is not sufficient, regardless of how powerful the considered models are in terms of parameter count. Finally, we discuss the recently revived idea of mixture of experts (MoE), a paradigm that implements routing-based inference for large language models. This paradigm is at the time of writing this book one of the main driving forces underlying existing, publicly available large language models. The technique considerably reduces the training and inference time, assuming adequate hardware.

We conclude the book by an overview of the main trends illustrated in the book and further directions. We observe that multimodal language models are becoming the dominant paradigm for multimodal machine learning and are nowadays capable of solving wide spectra of different tasks that entail both perception and reasoning. We finally discuss the hardware and algorithmic trends that arose due to increased scaling of multimodal models.

The purpose of this book is to illustrate some of the key techniques related to different modalities to an interested reader previously not familiar with them. The chapters include broad overviews and, however, are all equipped with "Method Spotlight" sections that dive deeper into technical specifics of selected methods. The purpose of these deep dives is to not only encourage the reader to dig deeper into the area of interest but also illustrate the type of mathematics required for understanding different paradigms – in most cases, linear algebra is a solid foundation.

Glossary

Artificial Intelligence (AI) A field of computer science focusing on understanding and building intelligent systems.

Machine Learning A subfield of AI that involves training algorithms to learn patterns and make predictions based on data without being explicitly programmed.

Classification A type of machine learning where an algorithm is trained to assign data to categories or classes based on its characteristics.

Regression A type of machine learning where an algorithm is trained to predict a continuous value based on input data.

Neural Network A type of machine learning algorithm that is modeled after the structure of the human brain and is used for tasks such as image recognition and natural language processing.

Deep Learning A type of machine learning that involves training neural networks with multiple layers to learn complex patterns and make predictions.

Supervised Learning A type of machine learning where the algorithm is trained on labeled data, meaning the desired output is already known.

Unsupervised Learning A type of machine learning where the algorithm is trained on unlabeled data, meaning the desired output is not known.

Modality A particular form or method of representing data, such as text, image, or audio.

Multimodality The use of multiple modalities, such as text, images, and audio, to represent data.

Feature Fusion (early fusion) The process of combining features extracted from each modality to create a joint feature representation.

Decision Fusion (late fusion) The process of combining decisions made by models trained on each modality to make a final decision.

Multimodal Embedding A joint embedding space that maps representations from different modalities into a common vector space.

Cross-Modal Retrieval The process of retrieving data from one modality based on a query in another modality, such as retrieving images based on a text query.

© The Author(s), under exclusive license to Springer Nature Switzerland AG 2024
B. Škrlj, *From Unimodal to Multimodal Machine Learning*, SpringerBriefs in
Computer Science, https://doi.org/10.1007/978-3-031-57016-2

Multimodal Machine Learning The process of training models to recognize and combine information from multiple modalities, such as text, images, and audio.

Unimodal Machine Learning The process of training models to recognize patterns within a single modality, such as text or images.

Graph Neural Networks A type of machine learning algorithm that enables deep learning for graph-based inputs (example includes considering the adjacency matrix and node features).

Large Language Models Branch of machine learning models mostly based on transformer architecture. They are trained to predict consequent tokens in text, and are capable of human-level comprehension and reasoning across many real-life benchmark tasks.

References

1. Alzahrani, Y., Boufama, B.: Biomedical image segmentation: a survey. SN Comput. Sci. **2**(4), 310 (2021). https://doi.org/10.1007/s42979-021-00704-7
2. Alzubi, J., Nayyar, A., Kumar, A.: Machine learning from theory to algorithms: an overview. In: Journal of Physics: Conference Series, vol. 1142, p. 012012. IOP Publishing (2018)
3. Arik, S.Ö., Pfister, T.: TabNet: attentive interpretable tabular learning. In: Thirty-Fifth AAAI Conference on Artificial Intelligence, AAAI 2021, Thirty-Third Conference on Innovative Applications of Artificial Intelligence, IAAI 2021, The Eleventh Symposium on Educational Advances in Artificial Intelligence, EAAI 2021, Virtual Event, 2–9 February 2021, pp. 6679–6687. AAAI Press (2021). https://doi.org/10.1609/aaai.v35i8.16826
4. Askari, A., Abolghasemi, A., Pasi, G., Kraaij, W., Verberne, S.: Injecting the BM25 score as text improves BERT-based re-rankers. In: Kamps, J., Goeuriot, L., Crestani, F., Maistro, M., Joho, H., Davis, B., Gurrin, C., Kruschwitz, U., Caputo, A. (eds.) Advances in Information Retrieval—45th European Conference on Information Retrieval, ECIR 2023, Dublin, Ireland, 2–6 April 2023, Proceedings, Part I, Lecture Notes in Computer Science, vol. 13980, pp. 66–83. Springer, Berlin (2023). https://doi.org/10.1007/978-3-031-28244-7_5
5. Ayodele, T.O.: Machine learning overview. New Adv. Mach. Learn. **2**, 9–18 (2010)
6. Alom, Md Zahangir, et al.: Recurrent residual convolutional neural network based on u-net (r2u-net) for medical image segmentation. arXiv preprint arXiv:1802.06955 (2018).
7. Baltrusaitis, T., Ahuja, C., Morency, L.: Multimodal Machine Learning: a survey and taxonomy. IEEE Trans. Pattern Anal. Mach. Intell. **41**(2), 423–443 (2019). https://doi.org/10.1109/TPAMI.2018.2798607
8. Bayoudh, K., Knani, R., Hamdaoui, F., Mtibaa, A.: A survey on deep multimodal learning for computer vision: advances, trends, applications, and datasets. Vis. Comput. **38**(8), 2939–2970 (2022). https://doi.org/10.1007/s00371-021-02166-7
9. Bengio, Y., Ducharme, R., Vincent, P.: A Neural probabilistic language model. In: Leen, T.K., Dietterich, T.G., Tresp, V. (eds.) Advances in Neural Information Processing Systems 13, Papers from Neural Information Processing Systems (NIPS) 2000, Denver, CO, USA, pp. 932–938. MIT Press, Cambridge (2000). https://proceedings.neurips.cc/paper/2000/hash/728f206c2a01bf572b5940d7d9a8fa4c-Abstract.html
10. Bojanowski, P., Grave, E., Joulin, A., Mikolov, T.: Enriching word vectors with subword information. Trans. Assoc. Comput. Linguist. **5**, 135–146 (2017). https://doi.org/10.1162/tacl_a_00051
11. Bowman, S.R., Angeli, G., Potts, C., Manning, C.D.: A large annotated corpus for learning natural language inference. In: Màrquez, L., Callison-Burch, C., Su, J., Pighin, D., Marton, Y. (eds.) Proceedings of the 2015 Conference on Empirical Methods in Natural Language

© The Author(s), under exclusive license to Springer Nature Switzerland AG 2024
B. Škrlj, *From Unimodal to Multimodal Machine Learning*, SpringerBriefs in Computer Science, https://doi.org/10.1007/978-3-031-57016-2

Processing, EMNLP 2015, Lisbon, Portugal, 17–21 September 2015, pp. 632–642. The Association for Computational Linguistics (2015). https://doi.org/10.18653/v1/d15-1075

12. Bregler, C., Covell, M., Slaney, M.: Video rewrite: driving visual speech with audio. In: Owen, G.S., Whitted, T., Mones-Hattal, B. (eds.) Proceedings of the 24th Annual Conference on Computer Graphics and Interactive Techniques, SIGGRAPH 1997, Los Angeles, CA, USA, 3–8 August 1997, pp. 353–360. ACM, New York (1997). https://doi.org/10.1145/258734.258880

13. Breiman, L.: Random forests. Mach. Learn. **45**(1), 5–32 (2001). https://doi.org/10.1023/A:1010933404324

14. Brown, T.B., Mann, B., Ryder, N., Subbiah, M., Kaplan, J., Dhariwal, P., Neelakantan, A., Shyam, P., Sastry, G., Askell, A., Agarwal, S., Herbert-Voss, A., Krueger, G., Henighan, T., Child, R., Ramesh, A., Ziegler, D.M., Wu, J., Winter, C., Hesse, C., Chen, M., Sigler, E., Litwin, M., Gray, S., Chess, B., Clark, J., Berner, C., McCandlish, S., Radford, A., Sutskever, I., Amodei, D.: Language models are few-shot learners. In: Larochelle, H., Ranzato, M., Hadsell, R., Balcan, M., Lin, H. (eds.) Advances in Neural Information Processing Systems 33: Annual Conference on Neural Information Processing Systems 2020, NeurIPS 2020, 6–12 December 2020, Virtual (2020). https://proceedings.neurips.cc/paper/2020/hash/1457c0d6bfcb4967418bfb8ac142f64a-Abstract.html

15. Cahyani, D.E., Patasik, I.: Performance comparison of TF-IDF and word2vec models for emotion text classification. Bull. Electr. Eng. Inform. **10**(5), 2780–2788 (2021)

16. Cao, Y., Steffey, S., He, J., Xiao, D., Tao, C., Chen, P., Müller, H.: Medical image retrieval: a multimodal approach. Cancer Inform. **13s3**, CIN.S14053 (2014). https://doi.org/10.4137/CIN.S14053. PMID: 26309389

17. Carion, N., Massa, F., Synnaeve, G., Usunier, N., Kirillov, A., Zagoruyko, S.: End-to-end object detection with transformers. In: Vedaldi, A., Bischof, H., Brox, T., Frahm, J. (eds.) Computer Vision—ECCV 2020—16th European Conference, Glasgow, UK, 23–28 August 2020, Proceedings, Part I, Lecture Notes in Computer Science, vol. 12346, pp. 213–229. Springer, Berlin (2020). https://doi.org/10.1007/978-3-030-58452-8_13

18. Chen, T.: Audiovisual speech processing. IEEE Signal Process. Mag. **18**(1), 9–21 (2001). https://doi.org/10.1109/79.911195

19. Chen, T., He, T., Benesty, M., Khotilovich, V., Tang, Y., Cho, H., Chen, K., Mitchell, R., Cano, I., Zhou, T., et al.: XGBoost: extreme gradient boosting. R package version 0.4-2 **1**(4), 1–4 (2015)

20. Chen, Y., Wei, F., Sun, X., Wu, Z., Lin, S.: A Simple multi-modality transfer learning baseline for sign language translation. In: IEEE/CVF Conference on Computer Vision and Pattern Recognition, CVPR 2022, New Orleans, LA, USA, 18–24 June 2022, pp. 5110–5120. IEEE, Piscataway (2022). https://doi.org/10.1109/CVPR52688.2022.00506

21. Cho, K., van Merrienboer, B., Gülçehre, Ç., Bahdanau, D., Bougares, F., Schwenk, H., Bengio, Y.: Learning phrase representations using RNN encoder-decoder for statistical machine translation. In: Moschitti, A., Pang, B., Daelemans, W. (eds.) Proceedings of the 2014 Conference on Empirical Methods in Natural Language Processing, EMNLP 2014, 25–29 October 2014, Doha, Qatar, A meeting of SIGDAT, a Special Interest Group of the ACL, pp. 1724–1734. ACL (2014). https://doi.org/10.3115/v1/d14-1179

22. Chua, T., Tang, J., Hong, R., Li, H., Luo, Z., Zheng, Y.: NUS-WIDE: a real-world web image database from National University of Singapore. In: Marchand-Maillet, S., Kompatsiaris, Y. (eds.) Proceedings of the 8th ACM International Conference on Image and Video Retrieval, CIVR 2009, Santorini Island, Greece, 8–10 July 2009. ACM, New York (2009). https://doi.org/10.1145/1646396.1646452

23. Ciresan, D.C., Meier, U., Masci, J., Gambardella, L.M., Schmidhuber, J.: Flexible, high performance convolutional neural networks for image classification. In: Walsh, T. (ed.) IJCAI 2011, Proceedings of the 22nd International Joint Conference on Artificial Intelligence, Barcelona, Catalonia, Spain, 16–22 July 2011, pp. 1237–1242. IJCAI/AAAI (2011). https://doi.org/10.5591/978-1-57735-516-8/IJCAI11-210

24. Collobert, R., Bengio, S., Mariéthoz, J.: Torch: a modular machine learning software library. Tech. rep., Idiap (2002)
25. Cover, T., Hart, P.: Nearest neighbor pattern classification. IEEE Trans. Inform. Theory **13**(1), 21–27 (1967)
26. Cui, P., Wang, X., Pei, J., Zhu, W.: A survey on network embedding. IEEE Trans. Knowl. Data Eng. **31**(5), 833–852 (2019). https://doi.org/10.1109/TKDE.2018.2849727
27. DeJong, G., Mooney, R.J.: Explanation-based learning: an alternative view. Mach. Learn. **1**(2), 145–176 (1986). https://doi.org/10.1023/A:1022898111663
28. Demsar, J., Curk, T., Erjavec, A., Gorup, C., Hocevar, T., Milutinovic, M., Mozina, M., Polajnar, M., Toplak, M., Staric, A., Stajdohar, M., Umek, L., Zagar, L., Zbontar, J., Zitnik, M., Zupan, B.: Orange: data mining toolbox in python. J. Mach. Learn. Res. **14**(1), 2349–2353 (2013). https://dl.acm.org/doi/10.5555/2567709.2567736
29. Deng, J., Dong, W., Socher, R., Li, L., Li, K., Fei-Fei, L.: ImageNet: a large-scale hierarchical image database. In: 2009 IEEE Computer Society Conference on Computer Vision and Pattern Recognition (CVPR 2009), Miami, Florida, USA, 20–25 June 2009, pp. 248–255. IEEE Computer Society (2009). https://doi.org/10.1109/CVPR.2009.5206848
30. Devlin, J., Chang, M., Lee, K., Toutanova, K.: BERT: pre-training of deep bidirectional transformers for language understanding. In: Burstein, J., Doran, C., Solorio, T. (eds.) Proceedings of the 2019 Conference of the North American Chapter of the Association for Computational Linguistics: Human Language Technologies, NAACL-HLT 2019, Minneapolis, MN, USA, 2–7 June 2019, vol. 1 (Long and Short Papers), pp. 4171–4186. Association for Computational Linguistics (2019). https://doi.org/10.18653/v1/n19-1423
31. Driess, D., Xia, F., Sajjadi, M.S.M., Lynch, C., Chowdhery, A., Ichter, B., Wahid, A., Tompson, J., Vuong, Q., Yu, T., Huang, W., Chebotar, Y., Sermanet, P., Duckworth, D., Levine, S., Vanhoucke, V., Hausman, K., Toussaint, M., Greff, K., Zeng, A., Mordatch, I., Florence, P.: PaLM-E: an embodied multimodal language model. In: Krause, A., Brunskill, E., Cho, K., Engelhardt, B., Sabato, S., Scarlett, J. (eds.) International Conference on Machine Learning, ICML 2023, 23–29 July 2023, Honolulu, Hawaii, USA, Proceedings of Machine Learning Research, vol. 202, pp. 8469–8488. PMLR (2023). https://proceedings.mlr.press/v202/driess23a.html
32. Du, N., Huang, Y., Dai, A.M., Tong, S., Lepikhin, D., Xu, Y., Krikun, M., Zhou, Y., Yu, A.W., Firat, O., Zoph, B., Fedus, L., Bosma, M.P., Zhou, Z., Wang, T., Wang, Y.E., Webster, K., Pellat, M., Robinson, K., Meier-Hellstern, K.S., Duke, T., Dixon, L., Zhang, K., Le, Q.V., Wu, Y., Chen, Z., Cui, C.: GLaM: efficient scaling of language models with mixture-of-experts. In: Chaudhuri, K., Jegelka, S., Song, L., Szepesvári, C., Niu, G., Sabato, S. (eds.) International Conference on Machine Learning, ICML 2022, 17–23 July 2022, Baltimore, Maryland, USA, Proceedings of Machine Learning Research, vol. 162, pp. 5547–5569. PMLR (2022). https://proceedings.mlr.press/v162/du22c.html
33. Dutoit, T.: An Introduction to Text-to-Speech Synthesis, vol. 3. Springer, Berlin (1997)
34. Evangelopoulos, G., Zlatintsi, A., Potamianos, A., Maragos, P., Rapantzikos, K., Skoumas, G., Avrithis, Y.: Multimodal saliency and fusion for movie summarization based on aural, visual, and textual attention. IEEE Trans. Multim. **15**(7), 1553–1568 (2013). https://doi.org/10.1109/TMM.2013.2267205
35. Farhadi, A., Hejrati, S.M.M., Sadeghi, M.A., Young, P., Rashtchian, C., Hockenmaier, J., Forsyth, D.A.: Every picture tells a story: generating sentences from images. In: Daniilidis, K., Maragos, P., Paragios, N. (eds.) Computer Vision—ECCV 2010, 11th European Conference on Computer Vision, Heraklion, Crete, Greece, 5–11 September 2010, Proceedings, Part IV, Lecture Notes in Computer Science, vol. 6314, pp. 15–29. Springer, Berlin (2010). https://doi.org/10.1007/978-3-642-15561-1_2
36. Feng, X., Jiang, Y., Yang, X., Du, M., Li, X.: Computer vision algorithms and hardware implementations: a survey. Integration **69**, 309–320 (2019). https://doi.org/10.1016/j.vlsi.2019.07.005

37. Fu, C., Chen, P., Shen, Y., Qin, Y., Zhang, M., Lin, X., Qiu, Z., Lin, W., Yang, J., Zheng, X., Li, K., Sun, X., Ji, R.: MME: a comprehensive evaluation benchmark for multimodal large language models. CoRR abs/2306.13394 (2023). https://doi.org/10.48550/arXiv.2306.13394

38. Gadzicki, K., Khamsehashari, R., Zetzsche, C.: Early vs late fusion in multimodal convolutional neural networks. In: IEEE 23rd International Conference on Information Fusion, FUSION 2020, Rustenburg, South Africa, 6–9 July 2020, pp. 1–6. IEEE, Piscataway (2020). https://doi.org/10.23919/FUSION45008.2020.9190246

39. Gallo, I., Calefati, A., Nawaz, S., Janjua, M.K.: Image and encoded text fusion for multimodal classification. In: 2018 Digital Image Computing: Techniques and Applications, DICTA 2018, Canberra, Australia, 10–13 December 2018, pp. 1–7. IEEE, Piscataway (2018). https://doi.org/10.1109/DICTA.2018.8615789

40. Gallo, I., Ria, G., Landro, N., Grassa, R.L.: Image and text fusion for UPMC food-101 using BERT and CNNs. In: 35th International Conference on Image and Vision Computing New Zealand, IVCNZ 2020, Wellington, New Zealand, 25–27 November 2020, pp. 1–6. IEEE, Piscataway (2020). https://doi.org/10.1109/IVCNZ51579.2020.9290622

41. Gavrilyuk, K., Sanford, R., Javan, M., Snoek, C.G.M.: Actor-transformers for group activity recognition. In: 2020 IEEE/CVF Conference on Computer Vision and Pattern Recognition, CVPR 2020, Seattle, WA, USA, 13–19 June 2020, pp. 836–845. Computer Vision Foundation/IEEE (2020). https://doi.org/10.1109/CVPR42600.2020.00092. https://openaccess.thecvf.com/content_CVPR_2020/html/Gavrilyuk_Actor-Transformers_for_Group_Activity_Recognition_CVPR_2020_paper.html

42. Goodfellow, I.J., Pouget-Abadie, J., Mirza, M., Xu, B., Warde-Farley, D., Ozair, S., Courville, A.C., Bengio, Y.: Generative adversarial nets. In: Ghahramani, Z., Welling, M., Cortes, C., Lawrence, N.D., Weinberger, K.Q. (eds.) Advances in Neural Information Processing Systems 27: Annual Conference on Neural Information Processing Systems 2014, Montreal, Quebec, Canada, 8–13 December 2014, pp. 2672–2680 (2014). https://proceedings.neurips.cc/paper/2014/hash/5ca3e9b122f61f8f06494c97b1afccf3-Abstract.html

43. Goodfellow, I.J., Pouget-Abadie, J., Mirza, M., Xu, B., Warde-Farley, D., Ozair, S., Courville, A.C., Bengio, Y.: Generative adversarial networks. Commun. ACM **63**(11), 139–144 (2020). https://doi.org/10.1145/3422622

44. Goyal, P., Ferrara, E.: Graph embedding techniques, applications, and performance: a survey. Knowl. Based Syst. **151**, 78–94 (2018). https://doi.org/10.1016/j.knosys.2018.03.022

45. Grave, E., Bojanowski, P., Gupta, P., Joulin, A., Mikolov, T.: Learning word vectors for 157 languages. In: Calzolari, N., Choukri, K., Cieri, C., Declerck, T., Goggi, S., Hasida, K., Isahara, H., Maegaard, B., Mariani, J., Mazo, H., Moreno, A., Odijk, J., Piperidis, S., Tokunaga, T. (eds.) Proceedings of the Eleventh International Conference on Language Resources and Evaluation, LREC 2018, Miyazaki, Japan, 7–12 May 2018. European Language Resources Association (ELRA) (2018). http://www.lrec-conf.org/proceedings/lrec2018/summaries/627.html

46. Grover, A., Leskovec, J.: node2vec: scalable feature learning for networks. In: Krishnapuram, B., Shah, M., Smola, A.J., Aggarwal, C.C., Shen, D., Rastogi, R. (eds.) Proceedings of the 22nd ACM SIGKDD International Conference on Knowledge Discovery and Data Mining, San Francisco, CA, USA, 13–17 August 2016, pp. 855–864. ACM, New York (2016). https://doi.org/10.1145/2939672.2939754

47. Guo, M., Xu, T., Liu, J., Liu, Z., Jiang, P., Mu, T., Zhang, S., Martin, R.R., Cheng, M., Hu, S.: Attention mechanisms in computer vision: a survey. Comput. Vis. Media **8**(3), 331–368 (2022). https://doi.org/10.1007/s41095-022-0271-y

48. Hamilton, W.L., Ying, Z., Leskovec, J.: Inductive representation learning on large graphs. In: Guyon, I., von Luxburg, U., Bengio, S., Wallach, H.M., Fergus, R., Vishwanathan, S.V.N., Garnett, R. (eds.) Advances in Neural Information Processing Systems 30: Annual Conference on Neural Information Processing Systems 2017, Long Beach, CA, USA, 4–9 December 2017, pp. 1024–1034 (2017). https://proceedings.neurips.cc/paper/2017/hash/5dd9db5e033da9c6fb5ba83c7a7ebea9-Abstract.html

49. He, K., Chen, X., Xie, S., Li, Y., Dollár, P., Girshick, R.B.: Masked autoencoders are scalable vision learners. In: IEEE/CVF Conference on Computer Vision and Pattern Recognition, CVPR 2022, New Orleans, LA, USA, 18–24 June 2022, pp. 15979–15988. IEEE, Piscataway (2022). https://doi.org/10.1109/CVPR52688.2022.01553

50. He, K., Zhang, X., Ren, S., Sun, J.: Deep residual learning for image recognition. In: 2016 IEEE Conference on Computer Vision and Pattern Recognition, CVPR 2016, Las Vegas, NV, USA, 27–30 June 2016, pp. 770–778. IEEE Computer Society (2016). https://doi.org/10.1109/CVPR.2016.90

51. Hearst, M.A., Dumais, S.T., Osuna, E., Platt, J., Scholkopf, B.: Support vector machines. IEEE Intell. Syst. Their Appl. **13**(4), 18–28 (1998)

52. Hebb, D.O.: The Organization of Behavior: A Neuropsychological Theory. Psychology Press, London (2005)

53. Ciresan, D.C., Meier, U., Masci, J., Gambardella, L.M., Schmidhuber, J.: Flexible, high performance convolutional neural networks for image classification. In: Walsh, T. (ed.) IJCAI 2011, Proceedings of the 22nd International Joint Conference on Artificial Intelligence, Barcelona, Catalonia, Spain, 16–22 July 2011, pp. 1237–1242. IJCAI/AAAI (2011). https://doi.org/10.5591/978-1-57735-516-8/IJCAI11-210

54. Hochreiter, S.: The vanishing gradient problem during learning recurrent neural nets and problem solutions. Int. J. Uncertain. Fuzziness Knowl. Based Syst. **6**(2), 107–116 (1998). https://doi.org/10.1142/S0218488598000094

55. Hochreiter, S., Schmidhuber, J.: Long short-term memory. Neural Comput. **9**(8), 1735–1780 (1997). https://doi.org/10.1162/neco.1997.9.8.1735

56. Jaafar, N., Lachiri, Z.: Multimodal fusion methods with deep neural networks and meta-information for aggression detection in surveillance. Expert Syst. Appl. **211**, 118523 (2023). https://doi.org/10.1016/j.eswa.2022.118523

57. Jacobs, R.A., Jordan, M.I., Nowlan, S.J., Hinton, G.E.: Adaptive mixtures of local experts. Neural Comput. **3**(1), 79–87 (1991)

58. Jaimes, A., Sebe, N.: Multimodal human-computer interaction: a survey. Comput. Vis. Image Underst. **108**(1–2), 116–134 (2007). https://doi.org/10.1016/j.cviu.2006.10.019

59. Jones, K.S.: Natural language processing: a historical review. In: Current Issues in Computational Linguistics: In Honour of Don Walker, pp. 3–16 (1994)

60. Joulin, A., Grave, E., Bojanowski, P., Mikolov, T.: Bag of tricks for efficient text classification. In: Lapata, M., Blunsom, P., Koller, A. (eds.) Proceedings of the 15th Conference of the European Chapter of the Association for Computational Linguistics, EACL 2017, Valencia, Spain, 3–7 April 2017, Volume 2: Short Papers, pp. 427–431. Association for Computational Linguistics (2017). https://doi.org/10.18653/v1/e17-2068

61. Jumper, J., Evans, R., Pritzel, A., Green, T., Figurnov, M., Ronneberger, O., Tunyasuvunakool, K., Bates, R., Žídek, A., Potapenko, A., et al.: Highly accurate protein structure prediction with AlphaFold. Nature **596**(7873), 583–589 (2021)

62. Karvelis, P.S., Gavrilis, D., Georgoulas, G.K., Stylios, C.D.: Topic recommendation using Doc2Vec. In: 2018 International Joint Conference on Neural Networks, IJCNN 2018, Rio de Janeiro, Brazil, July 8–13, 2018, pp. 1–6. IEEE, Piscataway (2018). https://doi.org/10.1109/IJCNN.2018.8489513

63. Kim, M., Rabelo, J., Okeke, K., Goebel, R.: Legal information retrieval and entailment based on BM25, transformer and semantic thesaurus methods. Rev. Socionetwork Strateg. **16**(1), 157–174 (2022). https://doi.org/10.1007/s12626-022-00103-1

64. Kingma, D.P., Welling, M.: Auto-encoding variational bayes. In: Bengio, Y., LeCun, Y. (eds.) 2nd International Conference on Learning Representations, ICLR 2014, Banff, AB, Canada, 14–16 April 2014, Conference Track Proceedings (2014). http://arxiv.org/abs/1312.6114

65. Kipf, T.N., Welling, M.: Semi-supervised classification with graph convolutional networks. In: 5th International Conference on Learning Representations, ICLR 2017, Toulon, France, 24–26 April 2017, Conference Track Proceedings. OpenReview.net (2017). https://openreview.net/forum?id=SJU4ayYgl

66. Kirchner, E.A., Fairclough, S.H., Kirchner, F.: Embedded multimodal interfaces in robotics: applications, future trends, and societal implications. In: Oviatt, S.L., Schuller, B.W., Cohen, P.R., Sonntag, D., Potamianos, G., Krüger, A. (eds.) The Handbook of Multimodal-Multisensor Interfaces: Language Processing, Software, Commercialization, and Emerging Directions—Volume 3. Association for Computing Machinery (2019). https://doi.org/10.1145/3233795.3233810

67. Kiros, R., Salakhutdinov, R., Zemel, R.S.: Multimodal neural language models. In: Proceedings of the 31th International Conference on Machine Learning, ICML 2014, Beijing, China, 21–26 June 2014, JMLR Workshop and Conference Proceedings, vol. 32, pp. 595–603. JMLR.org (2014). http://proceedings.mlr.press/v32/kiros14.html

68. Kiros, R., Salakhutdinov, R., Zemel, R.S.: Unifying visual-semantic embeddings with multimodal neural language models. CoRR abs/1411.2539 (2014). http://arxiv.org/abs/1411.2539

69. Klatt, D.H.: Review of text-to-speech conversion for English. J. Acoust. Soc. Am. **82**(3), 737–793 (1987)

70. Krizhevsky, A., Sutskever, I., Hinton, G.E.: ImageNet classification with deep convolutional neural networks. In: Bartlett, P.L., Pereira, F.C.N., Burges, C.J.C., Bottou, L., Weinberger, K.Q. (eds.) Advances in Neural Information Processing Systems 25: 26th Annual Conference on Neural Information Processing Systems 2012. Proceedings of a Meeting held 3–6 December 2012, Lake Tahoe, Nevada, United States, pp. 1106–1114 (2012). https://proceedings.neurips.cc/paper/2012/hash/c399862d3b9d6b76c8436e924a68c45b-Abstract.html

71. Kuhn, H.W.: The Hungarian method for the assignment problem. Naval Res Logist. Q. **2**(1–2), 83–97 (1955)

72. Kuipers, B., Feigenbaum, E.A., Hart, P.E., Nilsson, N.J.: Shakey: from conception to history. AI Mag. **38**(1), 88–103 (2017). https://doi.org/10.1609/aimag.v38i1.2716

73. Le, Q.V., Mikolov, T.: Distributed representations of sentences and documents. In: Proceedings of the 31th International Conference on Machine Learning, ICML 2014, Beijing, China, 21–26 June 2014, JMLR Workshop and Conference Proceedings, vol. 32, pp. 1188–1196. JMLR.org (2014). http://proceedings.mlr.press/v32/le14.html

74. LeCun, Y.: The MNIST database of handwritten digits (1998). http://yann.lecun.com/exdb/mnist/

75. LeCun, Y., Bottou, L., Bengio, Y., Haffner, P.: Gradient-based learning applied to document recognition. Proc. IEEE **86**(11), 2278–2324 (1998). https://doi.org/10.1109/5.726791

76. Li, S., Kulkarni, G., Berg, T.L., Berg, A.C., Choi, Y.: Composing simple image descriptions using web-scale n-grams. In: Goldwater, S., Manning, C.D. (eds.) Proceedings of the Fifteenth Conference on Computational Natural Language Learning, CoNLL 2011, Portland, Oregon, USA, 23–24 June 2011, pp. 220–228. ACL (2011). https://aclanthology.org/W11-0326/

77. Liang, P.P., Zadeh, A., Morency, L.: Foundations and recent trends in multimodal machine learning: principles, challenges, and open questions. CoRR abs/2209.03430 (2022). https://doi.org/10.48550/arXiv.2209.03430

78. Liao, L., He, X., Zhao, B., Ngo, C., Chua, T.: Interpretable multimodal retrieval for fashion products. In: Boll, S., Lee, K.M., Luo, J., Zhu, W., Byun, H., Chen, C.W., Lienhart, R., Mei, T. (eds.) 2018 ACM Multimedia Conference on Multimedia Conference, MM 2018, Seoul, Republic of Korea, 22–26 October 2018, pp. 1571–1579. ACM, New York (2018). https://doi.org/10.1145/3240508.3240646

79. Lin, T., Maire, M., Belongie, S.J., Hays, J., Perona, P., Ramanan, D., Dollár, P., Zitnick, C.L.: Microsoft COCO: common objects in context. In: Fleet, D.J., Pajdla, T., Schiele, B., Tuytelaars, T. (eds.) Computer Vision—ECCV 2014—13th European Conference, Zurich, Switzerland, 6–12 September 2014, Proceedings, Part V, Lecture Notes in Computer Science, vol. 8693, pp. 740–755. Springer, Berlin (2014). https://doi.org/10.1007/978-3-319-10602-1_48

80. Lin, T., Wang, Y., Liu, X., Qiu, X.: A survey of transformers. AI Open **3**, 111–132 (2022). https://doi.org/10.1016/j.aiopen.2022.10.001

81. Liu, A., Yuan, S., Zhang, C., Luo, C., Liao, Y., Bai, K., Xu, Z.: Multi-level multimodal transformer network for multimodal recipe comprehension. In: Huang, J.X., Chang, Y., Cheng, X. Kamps, J., Murdock, V., Wen, J., Liu, Y. (eds.) Proceedings of the 43rd International ACM SIGIR Conference on Research and Development in Information Retrieval, SIGIR 2020, Virtual Event, China, July 25–30, 2020, pp. 1781–1784. ACM, New York (2020). https://doi.org/10.1145/3397271.3401247

82. Luan, Y., Eisenstein, J., Toutanova, K., Collins, M.: Sparse, dense, and attentional representations for text retrieval. Trans. Assoc. Comput. Linguist. **9**, 329–345 (2021). https://doi.org/10.1162/tacl_a_00369

83. Martinc, M., Skrjanec, I., Zupan, K., Pollak, S.: PAN 2017: author profiling—gender and language variety prediction. In: Cappellato, L., Ferro, N., Goeuriot, L., Mandl, T. (eds.) Working Notes of CLEF 2017—Conference and Labs of the Evaluation Forum, Dublin, Ireland, 11–14 September 2017, CEUR Workshop Proceedings, vol. 1866. CEUR-WS.org (2017). https://ceur-ws.org/Vol-1866/paper_78.pdf

84. Masini, R.P., Medeiros, M.C., Mendes, E.F.: Machine learning advances for time series forecasting. J. Econ. Surv. **37**(1), 76–111 (2023)

85. McGurk, H., MacDonald, J.: Hearing lips and seeing voices. Nature **264**(5588), 746–748 (1976)

86. Michie, D., Chambers, R.A.: BOXES: An experiment in adaptive control. Mach. Intell. **2**(2), 137–152 (1968)

87. Mikolov, T., Sutskever, I., Chen, K., Corrado, G.S., Dean, J.: Distributed representations of words and phrases and their compositionality. In: Burges, C.J.C., Bottou, L., Ghahramani, Z., Weinberger, K.Q. (eds.) Advances in Neural Information Processing Systems 26: 27th Annual Conference on Neural Information Processing Systems 2013. Proceedings of a meeting held 5–8 December 2013, Lake Tahoe, Nevada, USA, pp. 3111–3119 (2013). https://proceedings.neurips.cc/paper/2013/hash/9aa42b31882ec039965f3c4923ce901b-Abstract.html

88. Min, B., Ross, H., Sulem, E., Veyseh, A.P.B., Nguyen, T.H., Sainz, O., Agirre, E., Heintz, I., Roth, D.: Recent advances in natural language processing via large pre-trained language models: a survey. ACM Comput. Surv. **56**(2), 30:1–30:40 (2024). https://doi.org/10.1145/3605943

89. Min, W., Jiang, S., Sang, J., Wang, H., Liu, X., Herranz, L.: Being a supercook: joint food attributes and multimodal content modeling for recipe retrieval and exploration. IEEE Trans. Multim. **19**(5), 1100–1113 (2017). https://doi.org/10.1109/TMM.2016.2639382

90. Minsky, M., Papert, S.A.: Perceptrons, Reissue of the 1988 Expanded Edition with a New Foreword by Léon Bottou: An Introduction to Computational Geometry. MIT Press, Cambridge (2017)

91. Morvant, E., Habrard, A., Ayache, S.: Majority vote of diverse classifiers for late fusion. In: Fränti, P., Brown, G., Loog, M., Escolano, F., Pelillo, M. (eds.) Structural, Syntactic, and Statistical Pattern Recognition—Joint IAPR International Workshop, S+SSPR 2014, Joensuu, Finland, 20–22 August 2014. Proceedings, Lecture Notes in Computer Science, vol. 8621, pp. 153–162. Springer, Berlin (2014). https://doi.org/10.1007/978-3-662-44415-3_16

92. Perozzi, B., Al-Rfou, R., Skiena, S.: DeepWalk: online learning of social representations. In: Macskassy, S.A., Perlich, C., Leskovec, J., Wang, W., Ghani, R. (eds.) The 20th ACM SIGKDD International Conference on Knowledge Discovery and Data Mining, KDD '14, New York, NY, USA, 24–27 August 2014, pp. 701–710. ACM, New York (2014). https://doi.org/10.1145/2623330.2623732

93. Priyanka, S.S., Kumar, T.K.: Multi-channel speech enhancement using early and late fusion convolutional neural networks. Signal Image Video Process. **17**(4), 973–979 (2023)

94. Prokhorenkova, L.O., Gusev, G., Vorobev, A., Dorogush, A.V., Gulin, A.: CatBoost: unbiased boosting with categorical features. In: Bengio, S., Wallach, H.M., Larochelle, H., Grauman, K., Cesa-Bianchi, N., Garnett, R. (eds.) Advances in Neural Information Processing Systems 31: Annual Conference on Neural Information Processing Systems 2018, NeurIPS 2018, 3–8 December 2018, Montréal, Canada, pp. 6639–6649 (2018). https://proceedings.neurips.cc/paper/2018/hash/14491b756b3a51daac41c24863285549-Abstract.html

95. Qiu, J., Dong, Y., Ma, H., Li, J., Wang, K., Tang, J.: Network embedding as matrix factorization: unifying deepwalk, LINE, PTE, and node2vec. In: Chang, Y., Zhai, C., Liu, Y., Maarek, Y. (eds.) Proceedings of the Eleventh ACM International Conference on Web Search and Data Mining, WSDM 2018, Marina Del Rey, CA, USA, 5–9 February 2018, pp. 459–467. ACM, New York (2018). https://doi.org/10.1145/3159652.3159706

96. Quinlan, J.R.: Induction of decision trees. Mach. Learn. **1**(1), 81–106 (1986). https://doi.org/10.1023/A:1022643204877

97. Quinlan, J.R.: C4. 5: Programs for Machine Learning. Elsevier, Amsterdam (2014)

98. Radford, A., Narasimhan, K., Salimans, T., Sutskever, I., et al.: Improving language understanding by generative pre-training (2018)

99. Rafailidis, D., Manolopoulou, S., Daras, P.: A unified framework for multimodal retrieval. Pattern Recogn. **46**(12), 3358–3370 (2013). https://doi.org/10.1016/j.patcog.2013.05.023. https://www.sciencedirect.com/science/article/pii/S0031320313002471

100. Redmon, J., Divvala, S.K., Girshick, R.B., Farhadi, A.: You only look once: unified, real-time object detection. In: 2016 IEEE Conference on Computer Vision and Pattern Recognition, CVPR 2016, Las Vegas, NV, USA, 27–30 June 2016, pp. 779–788. IEEE Computer Society (2016). https://doi.org/10.1109/CVPR.2016.91

101. Reed, S.E., Akata, Z., Yan, X., Logeswaran, L., Schiele, B., Lee, H.: Generative adversarial text to image synthesis. In: Balcan, M., Weinberger, K.Q. (eds.) Proceedings of the 33nd International Conference on Machine Learning, ICML 2016, New York City, NY, USA, 19–24 June 2016, JMLR Workshop and Conference Proceedings, vol. 48, pp. 1060–1069. JMLR.org (2016). http://proceedings.mlr.press/v48/reed16.html

102. Reimers, N., Gurevych, I.: Sentence-BERT: sentence embeddings using siamese BERT-networks. In: Proceedings of the 2019 Conference on Empirical Methods in Natural Language Processing and the 9th International Joint Conference on Natural Language Processing (EMNLP-IJCNLP), pp. 3982–3992. Association for Computational Linguistics, Hong Kong (2019). https://doi.org/10.18653/v1/D19-1410.

103. Ren, S., He, K., Girshick, R.B., Sun, J.: Faster R-CNN: towards real-time object detection with region proposal networks. In: Cortes, C., Lawrence, N.D., Lee, D.D., Sugiyama, M., Garnett, R. (eds.) Advances in Neural Information Processing Systems 28: Annual Conference on Neural Information Processing Systems 2015, Montreal, Quebec, Canada, 7–12 December 2015, pp. 91–99 (2015). https://proceedings.neurips.cc/paper/2015/hash/14bfa6bb14875e45bba028a21ed38046-Abstract.html

104. Ren, Y., Ruan, Y., Tan, X., Qin, T., Zhao, S., Zhao, Z., Liu, T.: Fastspeech: fast, robust and controllable text to speech. In: Wallach, H.M., Larochelle, H., Beygelzimer, A., d'Alché-Buc, F., Fox, E.B., Garnett, R. (eds.) Advances in Neural Information Processing Systems 32: Annual Conference on Neural Information Processing Systems 2019, NeurIPS 2019, December 8–14, 2019, Vancouver, BC, Canada, pp. 3165–3174 (2019). https://proceedings.neurips.cc/paper/2019/hash/f63f65b503e22cb970527f23c9ad7db1-Abstract.html

105. Rezatofighi, H., Tsoi, N., Gwak, J., Sadeghian, A., Reid, I.D., Savarese, S.: Generalized intersection over union: a metric and a loss for bounding box regression. In: IEEE Conference on Computer Vision and Pattern Recognition, CVPR 2019, Long Beach, CA, USA, 16–20 June 2019, pp. 658–666. Computer Vision Foundation/IEEE (2019). https://doi.org/10.1109/CVPR.2019.00075. http://openaccess.thecvf.com/content_CVPR_2019/html/Rezatofighi_Generalized_Intersection_Over_Union_A_Metric_and_a_Loss_for_CVPR_2019_paper.html

106. Ribeiro, L.F., Saverese, P.H., Figueiredo, D.R.: *Struc2vec*: learning node representations from structural identity. In: Proceedings of the 23rd ACM SIGKDD International Conference on Knowledge Discovery and Data Mining, KDD '17, pp. 385–394. Association for Computing Machinery, New York (2017). https://doi.org/10.1145/3097983.3098061

107. Riquelme, C., Puigcerver, J., Mustafa, B., Neumann, M., Jenatton, R., Pinto, A.S., Keysers, D., Houlsby, N.: Scaling vision with sparse mixture of experts. In: Ranzato, M., Beygelzimer, A., Dauphin, Y.N., Liang, P., Vaughan., J.W. (eds.) Advances in Neural Information Processing Systems 34: Annual Conference on Neural Information Processing Systems 2021,

NeurIPS 2021, 6–14 December 2021, virtual, pp. 8583–8595 (2021). https://proceedings.neurips.cc/paper/2021/hash/48237d9f2dea8c74c2a72126cf63d933-Abstract.html

108. Robertson, S.E., Zaragoza, H.: The probabilistic relevance framework: BM25 and beyond. Found. Trends Inf. Retr. **3**(4), 333–389 (2009). https://doi.org/10.1561/1500000019

109. Ronneberger, O., Fischer, P., Brox, T.: U-Net: convolutional networks for biomedical image segmentation. In: Navab, N., Hornegger, J., W.M.W. III, Frangi, A.F. (eds.) Medical Image Computing and Computer-Assisted Intervention—MICCAI 2015—18th International Conference Munich, Germany, 5–9 October 2015, Proceedings, Part III, Lecture Notes in Computer Science, vol. 9351, pp. 234–241. Springer, Berlin (2015). https://doi.org/10.1007/978-3-319-24574-4_28

110. Rosenblatt, F.: The perceptron: a probabilistic model for information storage and organization in the brain. Psychol. Rev. **65**(6), 386 (1958)

111. Ruder, S.: An overview of gradient descent optimization algorithms. CoRR abs/1609.04747 (2016). http://arxiv.org/abs/1609.04747

112. Salakhutdinov, R.: Deep learning. In: Macskassy, S.A., Perlich, C., Leskovec, J., Wang, W., Ghani, R. (eds.) The 20th ACM SIGKDD International Conference on Knowledge Discovery and Data Mining, KDD '14, New York, NY, USA, 24–27 August 2014, p. 1973. ACM, New York (2014). https://doi.org/10.1145/2623330.2630809

113. Salakhutdinov, R.: Deep learning. In: Macskassy, S.A., Perlich, C., Leskovec, J., Wang, W., Ghani, R. (eds.) The 20th ACM SIGKDD International Conference on Knowledge Discovery and Data Mining, KDD '14, New York, NY, USA, 24–27 August 2014, p. 1973. ACM, New York (2014). https://doi.org/10.1145/2623330.2630809

114. Salimans, T., Goodfellow, I.J., Zaremba, W., Cheung, V., Radford, A., Chen, X.: Improved techniques for training GANs. In: Lee, D.D., Sugiyama, M., von Luxburg, U., Guyon, I., Garnett, R. (eds.) Advances in Neural Information Processing Systems 29: Annual Conference on Neural Information Processing Systems 2016, Barcelona, Spain, 5–10 December 2016, pp. 2226–2234 (2016). https://proceedings.neurips.cc/paper/2016/hash/8a3363abe792db2d8761d6403605aeb7-Abstract.html

115. Samuel, A.L.: Some studies in machine learning using the game of checkers. IBM J. Res. Dev. **3**(3), 210–229 (1959)

116. Samuel, A.L.: Some studies in machine learning using the game of checkers. II–recent progress. IBM J. Res. Dev. **11**(6), 601–617 (1967)

117. Schuller, B.W., Valstar, M.F., Eyben, F., McKeown, G., Cowie, R., Pantic, M.: AVEC 2011-the first international audio/visual emotion challenge. In: D'Mello, S.K., Graesser, A.C., Schuller, B.W., Martin, J. (eds.) Affective Computing and Intelligent Interaction—Fourth International Conference, ACII 2011, Memphis, TN, USA, 9–12 October 2011, Proceedings, Part II, Lecture Notes in Computer Science, vol. 6975, pp. 415–424. Springer, Berlin (2011). https://doi.org/10.1007/978-3-642-24571-8_53

118. Sejnowski, T.J., Rosenberg, C.R.: Parallel networks that learn to pronounce English text. Complex Syst. **1**(1), 145–168 (1987)

119. Shazeer, N., Mirhoseini, A., Maziarz, K., Davis, A., Le, Q.V., Hinton, G.E., Dean, J.: Outrageously large neural networks: the sparsely-gated mixture-of-experts layer. In: 5th International Conference on Learning Representations, ICLR 2017, Toulon, France, 24–26 April 2017, Conference Track Proceedings. OpenReview.net (2017). https://openreview.net/forum?id=B1ckMDqlg

120. Silver, D., Hubert, T., Schrittwieser, J., Antonoglou, I., Lai, M., Guez, A., Lanctot, M., Sifre, L., Kumaran, D., Graepel, T., et al.: A general reinforcement learning algorithm that masters Chess, Shogi, and Go through self-play. Science **362**(6419), 1140–1144 (2018)

121. Simonyan, K., Zisserman, A.: Very deep convolutional networks for large-scale image recognition. In: Bengio, Y., LeCun, Y. (eds.) 3rd International Conference on Learning Representations, ICLR 2015, San Diego, CA, USA, 7–9 May 2015, Conference Track Proceedings (2015). http://arxiv.org/abs/1409.1556

122. Skrlj, B., Bevec, M., Lavrac, N.: Multimodal autoML via representation evolution. Mach. Learn. Knowl. Extr. **5**(1), 1–13 (2023). https://doi.org/10.3390/make5010001

123. Stahlschmidt, S.R., Ulfenborg, B., Synnergren, J.: Multimodal deep learning for biomedical data fusion: a review. Briefings Bioinform. **23**(2) (2022). https://doi.org/10.1093/bib/bbab569

124. Sun, J.W., Bao, J.Q., Bu, L.P.: Text classification algorithm Based on TF-IDF and BERT. In: 2022 11th International Conference of Information and Communication Technology (ICTech), pp. 1–4. IEEE, Piscataway (2022)

125. Szegedy, C., Liu, W., Jia, Y., Sermanet, P., Reed, S.E., Anguelov, D., Erhan, D., Vanhoucke, V., Rabinovich, A.: Going deeper with convolutions. In: IEEE Conference on Computer Vision and Pattern Recognition, CVPR 2015, Boston, MA, USA, 7–12 June 2015, pp. 1–9. IEEE Computer Society (2015). https://doi.org/10.1109/CVPR.2015.7298594

126. Tang, J., Qu, M., Mei, Q.: PTE: Predictive text embedding through large-scale heterogeneous text networks. In: Cao, L., Zhang, C., Joachims, T., Webb, G.I., Margineantu, D.D., Williams, G. (eds.) Proceedings of the 21th ACM SIGKDD International Conference on Knowledge Discovery and Data Mining, Sydney, NSW, Australia, 10–13 August 2015, pp. 1165–1174. ACM, New York (2015). https://doi.org/10.1145/2783258.2783307

127. Tang, J., Qu, M., Wang, M., Zhang, M., Yan, J., Mei, Q.: LINE: large-scale information network embedding. In: Gangemi, A., Leonardi, S., Panconesi, A. (eds.) Proceedings of the 24th International Conference on World Wide Web, WWW 2015, Florence, Italy, 18–22 May 2015, pp. 1067–1077. ACM, New York (2015). https://doi.org/10.1145/2736277.2741093

128. Taori, R., Gulrajani, I., Zhang, T., Dubois, Y., Li, X., Guestrin, C., Liang, P., Hashimoto, T.B.: Stanford Alpaca: an instruction-following LLaMA model (2023)

129. Thomason, J., Venugopalan, S., Guadarrama, S., Saenko, K., Mooney, R.J.: Integrating language and vision to generate natural language descriptions of videos in the wild. In: Hajic, J., Tsujii, J. (eds.) COLING 2014, 25th International Conference on Computational Linguistics, Proceedings of the Conference: Technical Papers, 23–29, Dublin, Ireland, August 2014, pp. 1218–1227. ACL (2014). https://aclanthology.org/C14-1115/

130. Touvron, H., Lavril, T., Izacard, G., Martinet, X., Lachaux, M., Lacroix, T., Rozière, B., Goyal, N., Hambro, E., Azhar, F., Rodriguez, A., Joulin, A., Grave, E., Lample, G.: LLaMA: Open and efficient foundation language models. CoRR abs/2302.13971 (2023). https://doi.org/10.48550/arXiv.2302.13971

131. Tsimpoukelli, M., Menick, J., Cabi, S., Eslami, S.M.A., Vinyals, O., Hill, F.: Multimodal few-shot learning with frozen language models. In: Ranzato, M., Beygelzimer, A., Dauphin, Y.N., Liang, P., Vaughan, J.W. (eds.) Advances in Neural Information Processing Systems 34: Annual Conference on Neural Information Processing Systems 2021, NeurIPS 2021, 6–14 December 2021, Virtual, pp. 200–212 (2021). https://proceedings.neurips.cc/paper/2021/hash/01b7575c38dac42f3cfb7d500438b875-Abstract.html

132. Turing, A.M.: Computing Machinery and Intelligence. Springer, Berlin (2009)

133. Vaswani, A., Shazeer, N., Parmar, N., Uszkoreit, J., Jones, L., Gomez, A.N., Kaiser, L., Polosukhin, I.: Attention is all you need. In: Guyon, I., von Luxburg, U., Bengio, S., Wallach, H.M., Fergus, R., Vishwanathan, S.V.N., Garnett, R. (eds.) Advances in Neural Information Processing Systems 30: Annual Conference on Neural Information Processing Systems 2017, Long Beach, CA, USA, 4–9 December 2017, pp. 5998–6008 (2017). https://proceedings.neurips.cc/paper/2017/hash/3f5ee243547dee91fbd053c1c4a845aa-Abstract.html

134. Velickovic, P., Cucurull, G., Casanova, A., Romero, A., Liò, P., Bengio, Y.: Graph attention networks. In: 6th International Conference on Learning Representations, ICLR 2018, Vancouver, BC, Canada, 30 April–3 May 2018, Conference Track Proceedings. OpenReview.net (2018). https://openreview.net/forum?id=rJXMpikCZ

135. Velickovic, P., Cucurull, G., Casanova, A., Romero, A., Liò, P., Bengio, Y.: Graph attention networks. In: 6th International Conference on Learning Representations, ICLR 2018, Vancouver, BC, Canada, 30 April–3 May 2018, Conference Track Proceedings. OpenReview.net (2018). https://openreview.net/forum?id=rJXMpikCZ

136. Wang, A., Pruksachatkun, Y., Nangia, N., Singh, A., Michael, J., Hill, F., Levy, O., Bowman, S.R.: SuperGLUE: a stickier Benchmark for general-purpose language understanding systems. In: Wallach, H.M., Larochelle, H., Beygelzimer, A., d'Alché-Buc, F., Fox, E.B., Garnett, R. (eds.) Advances in Neural Information Processing Systems 32: Annual Conference on Neural Information Processing Systems 2019, NeurIPS 2019, 8–14 December 2019, Vancouver, BC, Canada, pp. 3261–3275 (2019). https://proceedings.neurips.cc/paper/2019/hash/4496bf24afe7fab6f046bf4923da8de6-Abstract.html

137. Wang, A., Singh, A., Michael, J., Hill, F., Levy, O., Bowman, S.R.: GLUE: a multi-task benchmark and analysis platform for natural language understanding. In: 7th International Conference on Learning Representations, ICLR 2019, New Orleans, LA, USA, 6–9 May 2019. OpenReview.net (2019). https://openreview.net/forum?id=rJ4km2R5t7

138. Wang, X., Kumar, D., Thome, N., Cord, M., Precioso, F.: Recipe recognition with large multimodal food dataset. In: 2015 IEEE International Conference on Multimedia & Expo Workshops, ICME Workshops 2015, Turin, Italy, 29 June–3 July 2015, pp. 1–6. IEEE Computer Society (2015). https://doi.org/10.1109/ICMEW.2015.7169757

139. Watkins, C.J., Dayan, P.: Q-learning. Mach. Learn. **8**, 279–292 (1992)

140. Weizenbaum, J.: ELIZA—a computer program for the study of natural language communication between man and machine. Commun. ACM **9**(1), 36–45 (1966). https://doi.org/10.1145/365153.365168

141. Wen, Q., Zhou, T., Zhang, C., Chen, W., Ma, Z., Yan, J., Sun, L.: Transformers in time series: a survey. In: Proceedings of the Thirty-Second International Joint Conference on Artificial Intelligence, IJCAI 2023, 19–25 August 2023, Macao, SAR, China, pp. 6778–6786. ijcai.org (2023). https://doi.org/10.24963/ijcai.2023/759

142. Wolf, T., Debut, L., Sanh, V., Chaumond, J., Delangue, C., Moi, A., Cistac, P., Rault, T., Louf, R., Funtowicz, M., Brew, J.: Huggingface's transformers: state-of-the-art natural language processing. CoRR abs/1910.03771 (2019). http://arxiv.org/abs/1910.03771

143. Xia, F., Sun, K., Yu, S., Aziz, A., Wan, L., Pan, S., Liu, H.: Graph learning: a survey. IEEE Trans. Artif. Intell. **2**(2), 109–127 (2021). https://doi.org/10.1109/TAI.2021.3076021

144. Xiao, Y., Codevilla, F., Gurram, A., Urfalioglu, O., López, A.M.: Multimodal end-to-end autonomous driving. IEEE Trans. Intell. Transp. Syst. **23**(1), 537–547 (2022). https://doi.org/10.1109/TITS.2020.3013234

145. Xie, S., Girshick, R.B., Dollár, P., Tu, Z., He, K.: Aggregated residual transformations for deep neural networks. In: 2017 IEEE Conference on Computer Vision and Pattern Recognition, CVPR 2017, Honolulu, HI, USA, 21–26 July 2017, pp. 5987–5995. IEEE Computer Society (2017). https://doi.org/10.1109/CVPR.2017.634

146. Xu, K., Hu, W., Leskovec, J., Jegelka, S.: How powerful are graph neural networks? In: 7th International Conference on Learning Representations, ICLR 2019, New Orleans, LA, USA, 6–9 May 2019. OpenReview.net (2019). https://openreview.net/forum?id=ryGs6iA5Km

147. Xu, R., Xiong, C., Chen, W., Corso, J.J.: Jointly modeling deep video and compositional text to bridge vision and language in a unified framework. In: Bonet, B., Koenig, S. (eds.) Proceedings of the Twenty-Ninth AAAI Conference on Artificial Intelligence, Austin, Texas, USA, 25–30 January 2015, pp. 2346–2352. AAAI Press (2015). https://doi.org/10.1609/aaai.v29i1.9512

148. Yagcioglu, S., Erdem, A., Erdem, E., Ikizler-Cinbis, N.: RecipeQA: a challenge dataset for multimodal comprehension of cooking recipes. In: Proceedings of the 2018 Conference on Empirical Methods in Natural Language Processing, pp. 1358–1368. Association for Computational Linguistics, Brussels, Belgium (2018). https://aclanthology.org/D18-1166

149. Yawei, C., Min, C., Wenjing, G.: Multimodal taste classification of Chinese recipe based on image and text fusion. In: 2020 5th International Conference on Smart Grid and Electrical Automation (ICSGEA), pp. 203–208 (2020). https://doi.org/10.1109/ICSGEA51094.2020.00050

150. Yu, H., Wang, J., Huang, Z., Yang, Y., Xu, W.: Video paragraph captioning using hierarchical recurrent neural networks. In: 2016 IEEE Conference on Computer Vision and Pattern Recognition, CVPR 2016, Las Vegas, NV, USA, 27–30 June 2016, pp. 4584–4593. IEEE Computer Society (2016). https://doi.org/10.1109/CVPR.2016.496

151. Zadeh, A., Liang, P.P., Morency, L.: Foundations of multimodal co-learning. Inf. Fusion **64**, 188–193 (2020). https://doi.org/10.1016/j.inffus.2020.06.001

152. Zeroual, I., Lakhouaja, A.: Data science in light of natural language processing: an overview. Proc. Comput. Sci. **127**, 82–91 (2018)

153. Zhang, Y., Sidibé, D., Morel, O., Mériaudeau, F.: Deep multimodal fusion for semantic image segmentation: a survey. Image Vis. Comput. **105**, 104042 (2021). https://doi.org/10.1016/j.imavis.2020.104042

154. Zhou, J., Cui, G., Hu, S., Zhang, Z., Yang, C., Liu, Z., Wang, L., Li, C., Sun, M.: Graph neural networks: a review of methods and applications. AI Open **1**, 57–81 (2020). https://doi.org/10.1016/j.aiopen.2021.01.001

155. Zhu, X., Ghahramani, Z.: Learning from labeled and unlabeled data with label propagation. Tech. rep. (2002)

Printed in the United States
by Baker & Taylor Publisher Services